3181

QD
115
V4.7

ANALYSIS
WITH ION-SELECTIVE
ELECTRODES

ELLIS HORWOOD
SERIES IN ANALYTICAL CHEMISTRY

EDITORS: DR. R. A. CHALMERS & DR. MARY MASSON, University of Aberdeen

"I recommend that this Series be used as reference material. Its Authors are among the most respected in Europe". *J. Chemical Ed., New York.*

ANALYSIS WITH ION-SELECTIVE ELECTRODES

JOSEF VESELÝ
DALIBOR WEISS
Geological Survey, Prague

KAREL ŠTULÍK
Charles University, Prague

Translator:

MADELEINE ŠTULÍKOVÁ, BSc., MSc., PhD.

ELLIS HORWOOD LIMITED
Publishers • Chichester

Halsted Press: a division of
JOHN WILEY & SONS
New York • London • Sydney • Toronto

The publisher's colophon is reproduced from James Gillison's drawing of the ancient Market Cross, Chichester

First published in 1978 by
ELLIS HORWOOD LIMITED, PUBLISHERS
Market Cross House, Cooper Street, Chichester, Sussex, England

Distributors:

Australia, New Zealand, South-East Asia:

JOHN WILEY & SONS AUSTRALASIA PTY LIMITED
1-7 Waterloo Road, North Ryde, N. S. W., Australia

Canada:

JOHN WILEY & SONS CANADA LIMITED
22 Worcester Road, Rexdale, Ontario, Canada

Europe, Africa:

JOHN WILEY & SONS LIMITED
Baffins Lane, Chichester, Sussex, England

North and South America and the rest of the world:

HALSTED·PRESS a division of
JOHN WILEY & SONS INC.
605 Third Avenue, New York, N.Y. 10016, U.S.A.

© 1978 Josef Veselý, Dalibor Weiss, Karel Štulík/Ellis Horwood Ltd., Publishers

 British Library Cataloguing in Publication Data

Veselý, Josef
Analysis with ion-selective electrodes.
1. Chemistry, Analytic 2. Electrodes, Ion selective
I. Title II. Weiss, Dalibor
III. Štulík, Karel IV. Štulíková, Madeleine
543'.08   QD571   78-40226
ISBN 0-85312-092-7 (ELLIS HORWOOD)
ISBN 0-470-26296-6 (Halsted)

Printed in Great Britain by Biddles Ltd, Guildford, Surrey

AUTHORS' PREFACE

Ion-selective electrodes have become one of the most useful tools for rapid analysis or on-line monitoring in contemporary analytical chemistry. As such, they have already been treated in a number of reviews and monographs although no book has been devoted to analytical aspects.* The collection of lectures, *Ion-Selective Electrodes*, edited by R. A. Durst and published by the National Bureau of Standards, Washington 1969, is a classic in the field but is already outdated. The book by J. Koryta, *Ion-Selective Electrodes*, Cambridge University Press 1975, is directed primarily to the physico-chemical aspects of membrane potentials, but also gives a brief survey of analytical applications. Two other books, namely, that by G. J. Moody and J. D. R. Thomas, *Selective Ion-Sensitive Electrodes,* Merrow, Watford 1971, and that by K. Camman, *Das Arbeiten mit ion selektiven Elektroden,* Springer Verlag, Berlin 1973, (2nd Ed., 1977) deal to a greater or lesser degree with various aspects of the topic. Recently, more books have appeared, e.g. N. Lakshminarayanaiah, *Membrane Electrodes,* Academic Press, New York 1976, and P. L. Bailey, *Analysis with Ion-Selective Electrodes,* Heyden, London 1976, and many reviews dealing with various aspects of the field.

The present book attempts a general up-to-date review of the field and places the greatest emphasis upon analytical use of the electrodes; consequently, apparatus and measuring techniques are treated in detail, while the theoretical part gives a survey of the results obtained, without detailed derivation of the pertinent equations. Important analytical applications are discussed in detail and groups of similar determinations are tabulated. No detailed treatment of pH-measurement is given, as this field has been covered by good monographs and reviews.

We are indebted to all the authors, publishers and manufacturers who gave us permission to use their material in the book. Further we would like to thank Dr. M. Štulíková for helpful comments on the text during translation of the Czech manuscript, and Mr. J. Šunka for help in the preparation of the figures.

---

* Since preparation of this book, a volume devoted to applications in organic analysis has appeared in this series (see Baiulescu and Coşofreţ in list of titles available).

TABLE OF CONTENTS

10

LIST OF IMPORTANT SYMBOLS

a	activity	U	particle mobility
c, C	concentration	V	solution volume
D	diffusion coefficient	z	electric charge on a particle
E	electrode potential	α	side-reaction coefficient;
E^0	standard potential		degree of dissociation
$E^{0'}$	normal potential	γ	activity coefficient
F	Faraday constant	λ	equivalent conductivity
h	hydration number	Λ	equivalent conductivity at
I	ionic strength		a given concentration
J	material flux	μ	stoichiometric coefficient;
k	distribution coefficient		chemical potential
k_{IJ}^{Pot}	selectivity coefficient	μ^0	standard chemical potential
K	equilibrium constant	$\tilde{\mu}$	electrochemical potential
K_s	solubility product	φ	internal electrical potential
m	molality	$\Delta\varphi$	potential difference
R	universal gas constant	$\Delta\varphi_D$	Donnan potential
S	slope of the dependence of E vs. log a	$\Delta\varphi_L$	liquid-junction potential
		$\Delta\varphi_M$	membrane potential
t	transport number	Φ	osmotic coefficient
T	absolute temperature		

Chapter 1

INTRODUCTION AND THEORY

The oldest and still by far the best ion-selective electrode (ISE) is the glass electrode for pH measurements, which was discovered at the beginning of the present century [1, 2], and became a standard laboratory tool in the late thirties. Attempts to prepare glass electrodes selective for other ions appeared rather early [3, 4] and at present there are commercially available glass electrodes selective for several cations, chiefly the alkali metals and silver (see e.g. [5]). Some early experiments with solid-membrane ISE's were largely unsuccessful (see e.g. [6 – 11]). Rapid development in this field was started in the sixties by construction of working heterogeneous-membrane electrodes [12] and especially by the discovery of the single-crystal fluoride-selective electrode [13]. The field of ISE's was broadened by introduction of liquid ion-exchanger membranes [14, 15], membranes containing electroneutral macrocyclic compounds [16], enzyme electrodes [17] and gas sensors [134]. At present, many kinds of ion can be determined directly and many compounds indirectly with ISE's. Several dozens of manufacturers all over the world make these electrodes and many ISE's can readily be prepared in the laboratory. The main appeal of ISE's lies in the simplicity of the measuring technique and instrumentation and in their suitability for continuous monitoring, which makes them particularly useful in routine control analysis, pollution control and biology and medicine.

1.1 ION-SELECTIVE ELECTRODE POTENTIAL

When two phases containing electrically charged particles (ions, electrons or dipoles) come into contact, an electrical potential difference develops at their interface. The compositions of the phases are characterized by the appropriate electrochemical potentials, $\tilde{\mu}$, the values of which must be equal if the system is in equilibrium. Thus for the ith charged species present in phases 1 and 2,

$$\tilde{\mu}_i(1) = \tilde{\mu}_i(2) \tag{1.1}$$

Under ISE conditions it can be assumed that the chemical properties of the phases are independent of the electrical charges involved and thus Eq. (1.1) can be rewritten in terms of the standard chemical potentials, μ_i^0, the activities, a_i, the charges, z_i, and the inner electrical potentials of the phases, φ:

$$\mu_i^0(1) + RT \ln a_i(1) + z_i F\varphi(1) = \mu_i^0(2) + RT \ln a_i(2) + z_i F\varphi(2) \quad (1.2)$$

where R is the universal gas constant, T is the absolute temperature and F is the Faraday. Hence, for the potential difference at the interface,

$$\Delta\varphi = \varphi(2) - \varphi(1) = \frac{\mu_i^0(1) - \mu_i^0(2)}{z_i F} + \frac{RT}{z_i F} \ln \frac{a_i(1)}{a_i(2)}$$

$$= \text{const} + \frac{RT}{z_i F} \ln \frac{a_i(1)}{a_i(2)} \quad (1.3)$$

Therefore, if the activity of the ith species in one phase and the μ_i^0 values are known, the activity of the ith species in the other phase can be determined by measuring the $\Delta\varphi$ value. It can be seen that an immense number of systems can potentially be used as electrodes for determining the activities of various species in solution. However, in practical measurements not only sensitivity to the particular species, but also satisfactory selectivity, accuracy and reproducibility of the measurement will be required, thus imposing considerable limitations on the selection of useful systems.

First, most substances exhibiting purely electronic conductivity (chiefly metals, various forms of carbon and some other substances) are excluded as potential ion-selective materials, as they behave as redox electrodes and are sensitive to redox systems and not to individual ions. Substances suitable for use as ion-selective materials therefore exhibit ionic conductivity; in some cases both ions and electrons participate in the transport of electric charge in the material and whether the material will exhibit predominantly ionic or redox sensitivity depends chiefly on the relative values of the transport numbers for the ions and electrons. For example, in the series of chalcogenides the materials have progressively more metallic character in the order PbS < PbSe < PbTe, and the sensitivity towards Pb^{2+} decreases accordingly [18].

There are several exceptions where substances with predominantly electronic conductivity exhibit ion-sensitivity and these will be discussed in Section 1.1.2.

Second, the material must differentiate among various ionic species; ideally it should respond to only one of them. This is achieved, e.g. when various ionic species have different mobilities in the material and/or the material behaves as an ion-exchanger, an adsorbent, an extraction system or a complexing agent, selective toward certain ionic species.

The potential difference $\Delta\varphi$, given by Eq. (1.3), cannot be measured directly. Generally, only the potential difference in an electrochemical cell consisting of two electrodes (half-cells) is measurable. Therefore, the ion-selective material must be incorporated into some kind of electrode, the potential of which is then measured against a suitable reference electrode (direct potentiometric measurement). It is possible to connect the ion-selective substance directly to an electronic conductor (e.g. by depositing it on a carbon support or by pressing or plating a metal contact onto it); this principle is, in fact, employed in some types of ISE's (see Section 2.1.3). However, this arrangement, though advantageous from the point of view of electrode construction, has two disadvantages from the theoretical point of view. First, the potential difference formed at the interface of the ion-sensitive material and the electronic conductor is mostly poorly defined thermodynamically and may sometimes be poorly reproducible; second, the activity of the species sensed in the electrode active material is often not defined. The situation is simpler when the ion-sensitive material is placed between the sample solution and a standard solution containing a suitable reference electrode and the test species at defined activity. The ion-sensitive material then acts as an electrochemical membrane separating the sample and the standard solution and the potential difference between two reference electrodes, one immersed in the test solution and the other in the standard solution, is measured. The electrochemical cell employed can then be represented e.g. by the scheme

$$Hg\,|\,Hg_2Cl_2\,|\,sat.\,KCl\,||\,test\,|\quad membrane \quad|\,standard\,||\,sat.\,KCl\,|\,Hg_2Cl_2\,|\,Hg$$

$$\qquad\qquad\qquad\qquad solution \qquad\qquad\qquad solution$$

$$\qquad\qquad\qquad\qquad phase\ 1 \qquad\qquad\qquad\quad phase\ 2 \qquad\qquad\qquad (I)$$

$$\Delta\varphi_L(1)\ \varphi(1)\quad \Delta\varphi_M\quad \varphi(2)\ \Delta\varphi_L(2)$$

For this system Eq. (1.3) is valid, potentials $\varphi(1)$ and $\varphi(2)$ correspond to the test and the standard solution, respectively, and their difference is equal to the membrane potential, $\Delta\varphi_M$. With ion-selective membranes, the bulk of the membrane and of the solutions is often not in equilibrium, and mass transport then occurs; if at least the interfaces are in equilibrium, then Eq. (1.3) still applies, but the activity and potential values refer to the close vicinity of the interfaces.

It can be seen from scheme (I) that the actual potential difference measured will involve the membrane potential, $\Delta\varphi_M$, and liquid-junction potentials at the liquid phase interfaces. Potential $\Delta\varphi_L(2)$ need not always be present, as for example in the cell:

Hg | Hg₂Cl₂ | sat. KCl | | test | membrane | standard | AgCl | Ag
                          solution   solution
                                                      + 0.1M KCl

$$\underbrace{}_{\Delta\varphi_L(1)}\qquad\underbrace{}_{\Delta\varphi_M}$$

Cell type (II) is most frequently employed in measurements with ISE's. Nevertheless, it is now necessary to examine the properties of these potentials.

1.1.1 Theoretical Membrane Potentials

The membrane can simply mechanically prevent rapid mixing of two solutions, without hindering the diffusional transport of any species across the interface. Owing to the various diffusion rates of the individual ions, a potential gradient develops across the interface and gives rise to a potential difference between the two solutions, termed the liquid-junction or diffusion potential, $\Delta\varphi_L$. This is a limiting case with membrane potentials; in practical electrochemical cells containing ISE's it appears at the porous diaphragm of the external reference electrodes and the diffusion potential across the membrane often contributes to the overall membrane potential.

The value of the diffusion potential can be calculated by starting from the Nernst–Planck equations for the mass fluxes of all the species involved [19−22]; for the ith species travelling along the x-axis, assuming low concentration so that $c_i = a_i$, and the absence of convectional movement,

$$J_i = -U_i RT(dc_i/dx) - z_i F U_i c_i(d\varphi/dx) \qquad (1.4)$$

where U_i is the species mobility, defined by

$$U_i = \lambda_i/(|z_i|F^2) \qquad (1.5)$$

where λ_i is the equivalent ionic conductivity. On introducing the condition that no net electric current flows across the interface, and rearranging, the equation for the liquid-junction potential is obtained in the form

$$\Delta\varphi_L = \varphi_2 - \varphi_1 = -\frac{RT}{F}\int_1^2 \sum_i \frac{t_i}{z_i} d\ln c_i \qquad (1.6)$$

where t_i is the transport number of the ith species. In order to evaluate the integral in Eq. (1.6), the concentrations and the transport numbers must be known as functions of x. An approximate solution has been given by Henderson [23, 24] assuming that the activity coefficients of the species are equal to unity across the interface, i.e. $a_i = c_i$, and that the concentration gradient along the x-coordinate is linear (continuous mixture type of liquid junction — see Section 2.3.2). Then, for a liquid junction of thickness d, $[d = x(d) - x(0)]$,

$$\Delta\varphi_L = -\frac{RT}{F} \frac{\sum_i z_i U_i [c_i(d) - c_i(0)]}{\sum_i z_i^2 U_i [c_i(d) - c_i(0)]} \ln \frac{\sum_i z_i^2 U_i c_i(d)}{\sum_i z_i^2 U_i c_i(0)} \qquad (1.7)$$

The Henderson equation is most frequently employed for approximate determination of liquid-junction potentials; the error can be several millivolts (see Section 2.3.4). A more general solution of Eq. (1.6) has been given by Planck [21, 22]; however, these equations are used less frequently, as they do not yield the liquid-junction potential value explicitly. In two cases the liquid-junction potential can be calculated simply. First, if the liquid junction separates two solutions of a single electrolyte at concentrations c_1 and c_2, the liquid-junction potential is independent of the concentration distribution across the interface and is given by [25]:

$$\Delta\varphi_L = -\frac{RT}{F} \left(\frac{t_+}{z_+} + \frac{t_-}{z_-} \right) \ln \frac{c_2}{c_1} \qquad (1.8)$$

where subscripts $+$ and $-$ refer to the cation and anion, respectively. Second, when the liquid junction separates two uni-univalent electrolytes 1 and 2 with a common cation or anion present in equal concentrations, then the Henderson formula is simplified to give [26]:

$$\Delta\varphi_L = \pm \frac{RT}{F} \ln \frac{\Lambda(2)}{\Lambda(1)} \qquad (1.9)$$

where $\Lambda(1)$ and $\Lambda(2)$ are the equivalent conductivities at the given concentration, the $+$ sign holding for solutions with a common cation and the $-$ sign for solutions with a common anion. Recently, several authors have taken a somewhat different approach to the calculation of liquid-junction potentials (see e.g. [27]).

Of course, the conditions in practical measurements rarely permit uhfficiently accurate calculation of liquid-junction potential values. It is suerefore necessary to minimize their value or at least keep it constant dtring measurements. This is usually done by adding sufficiently high concentrations of indifferent electrolytes to the test solution (see Section 2.3.3).

The other limiting situation occurs when the membrane separating two solutions completely prevents transport of one or more charged species across the boundary. Then the ions to which the membrane is permeable will pass from the solution of higher concentration into the other, thus generating a potential difference between the two solutions and giving rise to an electrical double-layer at the two boundaries of the membrane. For example, for ions X^+ and Y^-, one of which passes through the membrane and the other does not, at equilibrium

$$\tilde{\mu}_{X^+}(1) = \tilde{\mu}_{X^+}(2)$$
$$\tilde{\mu}_{Y^-}(1) = \tilde{\mu}_{Y^-}(2)$$

(1.10)

and consequently

$$\Delta\varphi_D = \varphi_2 - \varphi_1 = \frac{RT}{F}\ln\frac{a_{x^+}(1)}{a_{x^+}(2)} = \frac{RT}{F}\ln\frac{a_{Y^-}(2)}{a_{Y^-}(1)}$$

(1.11)

The $\Delta\varphi_D$ value is termed the Donnan potential [28].

The properties of most membranes will lie between the two above-mentioned extremes. The passage of various ions will be hindered to various degrees by the effect of pore size and/or charges of opposite sign fixed on the pore walls. Typical examples of this type of membrane are ion-exchangers. The theory for these permselective membranes was developed chiefly by Sollner [29, 30], Teorell [31 − 35] and Meyer and Sievers [36, 37]. The basic concept is the assumption that the overall potential difference across the membrane is the sum of the two potential differences at the membrane–solution interfaces and the diffusion potential across the membrane. The resulting equations for $\Delta\varphi_M$ then consist of a combination of the Donnan and Henderson (Planck) terms. The equations for the potential of ion-selective membrane electrodes have been derived on the basis of this concept. For this purpose, the ISE types have been classified as those with solid membranes with fixed ion-exchange sites, liquid membranes with dissolved ion-exchangers and liquid membranes containing electroneutral ion-carriers; they will be treated in the following sections.

1.1.1.1 MEMBRANES WITH FIXED ION-EXCHANGE SITES

The theory for this group of electrodes has been the most thoroughly developed of all ISE theories, because the glass electrode, studied for several decades, belongs in this group. The foundations of this theory were laid by Nikolskii [38], on the basis of the ion-exchange concept of Horowitz [39] and Schiller [40]. Nikolskii assumed that the activities of the ions in the membrane were equal to the concentrations and that the total concentration of ions, c_0, was constant and equal to the number of ion-exchange sites. Then, when ions I^{z_1+} and J^{z_2+} are present in the solution, the exchange reaction

$$I^{z_1+}(s) + J^{z_2+}(l) \overset{K_{eq}}{\rightleftharpoons} I^{z_1+}(l) + J^{z_2+}(s) \qquad (1.12)$$

where (s) and (l) refer to the solid and liquid phase, respectively, takes place on the ion-exchange sites in the membrane, with equilibrium constant K_{eq}. The interface potential is then given by

$$\Delta\varphi = \Delta\varphi_0 + \frac{RT}{z_1 F}\ln\left[a_{I^{z_1+}} + K_{eq}(a_{J^{z_2+}})^{z_1/z_2}\right] \qquad (1.13)$$

where $\Delta\varphi_0 = \Delta\varphi_0' - (RT)/(z_1 F)\ln c_0$. The values calculated from Eq. (1.13) were often at variance with the experimental results, and Nikolskii and co-workers [41−44] later developed an ion-exchange theory for membrane materials with several kinds of ion-exchange sites with different activities, which gave a better fit to the experimental data. However, these treatments did not consider the existence of a diffusion potential across the membrane.

A general solution, taking into account both the ion-exchange equilibrium and the diffusion potential in the membrane, was given by Eisenman and co-workers [45−48]. For equilibrium (1.12), where $z_1 = z_2 = 1$, the membrane potential is given by

$$\Delta\varphi_M = \frac{RT}{F}\ln\frac{a_I(1) + k_{IJ}^{Pot}a_J(1)}{a_I(2) + k_{IJ}^{Pot}a_J(2)} \qquad (1.14)$$

or

$$\Delta\varphi_M = \frac{RT}{F}\ln\frac{a_J(1) + k_{JI}^{Pot}a_I(1)}{a_J(2) + k_{JI}^{Pot}a_I(2)} \qquad (1.15)$$

where (1) and (2) refer to the solutions on the two sides of the membrane. Equations (1.14) and (1.15) are similar to the Nikolskii equation (1.13), but in place of the ion-exchange equilibrium constant, K_{eq}, have the quantity k_{IJ}^{Pot} or k_{JI}^{Pot}, defined by

$$k_{IJ}^{Pot} = \frac{U_J}{U_I}K_{eq} \quad \text{or} \quad k_{JI}^{Pot} = \frac{U_I}{U_J K_{eq}}; \qquad k_{IJ}^{Pot} = \frac{1}{k_{JI}^{Pot}} \qquad (1.16)$$

where U_I and U_J are the mobilities of ions I and J, respectively, in the membrane and k^{Pot} is the selectivity coefficient, a very important value for practical use, indicating the degree of selectivity of the electrode for one kind of ions with respect to the other. If, say, $a_I(1) \gg k_{IJ}^{Pot} a_J(1)$, the electrode is selective to ion I and its potential depends on a_I with a slope of 59 mV per decade (the Nernstian slope). On the other hand, if $a_I(1) \ll k_{IJ}^{Pot} a_J(1)$, i.e. $a_J(1) \gg k_{JI}^{Pot} a_I(1)$, the electrode is selective for ion J. Equations (1.14) and (1.15) can be generalized for n ions in the solution with various charges of the same sign, z_i, in the form

$$\Delta \varphi_M = \frac{RT}{z_1 F} \ln \frac{a_1(1) + \sum\limits_{i=2}^{n} k_{1i}^{Pot} a_i(1)^{z_1/z_i}}{a_1(2) + \sum\limits_{i=2}^{n} k_{1i}^{Pot} a_i(2)^{z_1/z_i}} \qquad (1.17)$$

The equations above were obtained on the assumption of ideal behaviour of the ions in the membrane, i.e. $a_i = c_i$. Eisenman [48] also derived an equation for the case when only two ions are present but behave non-ideally to a certain degree, namely, the relationship between their activities and concentrations is given by

$$d \ln a_I / d \ln c_I = d \ln a_J / d \ln c_J = n; \qquad dn/dx = 0 \qquad (1.18)$$

The membrane potential for this so-called "n-type non-ideality" is then

$$\Delta \varphi_M = \frac{nRT}{F} \ln \frac{a_I(1)^{1/n} + (k_{IJ}^{Pot})^{1/n} a_J(1)^{1/n}}{a_I(2)^{1/n} + (k_{IJ}^{Pot})^{1/n} a_J(2)^{1/n}} \qquad (1.19)$$

1.1.1.2 LIQUID MEMBRANES WITH DISSOLVED ION-EXCHANGERS

These membranes consist of an inert organic solvent, containing a dissolved ion-exchanger, which selectively associates with certain ions present in solutions 1 and 2 separated by the membrane. The ion-exchanger is insoluble in solutions 1 and 2. Therefore, the main difference from the fixed site membranes is the free mobility of the ion-exchange sites within the membrane. A theory of membrane potentials under these conditions, assuming ideal behaviour of the system, has been given by Eisenman and co-workers [48–53]. For two ions, I^+ and J^+, present in solutions 1 and 2 and associating with ion-exchange sites S^- to give IS and JS with the association constants

$$K_{IS} = \frac{c_{IS}}{c_I c_S}; \qquad K_{JS} = \frac{c_{JS}}{c_J c_S} \qquad (1.20)$$

and the distribution coefficients for partition between the solution and the membrane, k_I and k_J, given by

$$k_I = \exp\left[(\mu_I^0(l) - \mu_I^0(m))/RT\right]$$
$$k_J = \exp\left[(\mu_J^0(l) - \mu_J^0(m))/RT\right]$$

(1.21)

where (l) and (m) refer to the solution and the membrane, respectively, it holds that

$$\frac{a_I(1)\,k_I}{c_I(0)} = \frac{a_J(1)\,k_J}{c_J(0)}$$

(1.22)

where $c(0)$ denotes the concentration at a boundary of a membrane of thickness $d = x(d) - x(0)$. The equations for the membrane potential can be obtained for two limiting cases, namely, for a virtually completely dissociated system, i.e. $c_{IS} \to 0$, $c_{JS} \to 0$,

$$\Delta\varphi_M = \frac{RT}{F}\ln\frac{a_I(1) + \dfrac{U_J}{U_I}\cdot\dfrac{k_J}{k_I}\,a_J(1)}{a_I(2) + \dfrac{U_J}{U_I}\cdot\dfrac{k_J}{k_I}\,a_J(2)}$$

(1.23)

and for a strongly associated system, i.e. $c_{IS} \gg c_S$, $c_{JS} \gg c_S$

$$\Delta\varphi_M = \frac{RT}{F}\left\{(1-\tau)\ln\frac{a_I(1) + \left[\dfrac{(U_J + U_S)\,k_J}{(U_I + U_S)\,k_I}\right]a_J(1)}{a_I(2) + \left[\dfrac{(U_J + U_S)\,k_J}{(U_I + U_S)\,k_I}\right]a_J(2)}\right.$$

$$\left. + \tau\ln\frac{a_I(1) + \left[\dfrac{U_{JS}}{U_{IS}}K_{IJ}\right]a_J(1)}{a_I(2) + \left[\dfrac{U_{JS}}{U_{IS}}K_{IJ}\right]a_J(2)}\right\}$$

(1.24)

where K_{IJ} is the ion-exchange constant of the reaction

$$I(m) + J(s) \rightleftharpoons I(s) + J(m)$$

and given by

$$K_{IJ} = \frac{k_J}{k_I}\cdot\frac{K_{JS}}{K_{IS}}$$

(1.25)

and τ is given by

$$\tau = \frac{U_S(U_{JS}K_{JS} - U_{IS}K_{IS})}{(U_I + U_S)\,U_{JS}K_{JS} - (U_J + U_S)\,U_{IS}K_{IS}}$$

(1.26)

While Eq. (1.23) can readily be rewritten in terms of the selectivity coefficient, where

$$k_{IJ}^{Pot} = \frac{U_J}{U_I} \cdot \frac{k_J}{k_I} \qquad (1.27)$$

Eq. (1.24) does not immediately yield an explicit form for this important quantity. However, assuming that either the exchange sites move very rapidly in the membrane and no diffusion potential is formed across the membrane, $\tau \to 1$, or the exchange sites are very poorly mobile, $\tau \to 0$, Eq. (1.24) simplifies to give

$$k_{IJ}^{Pot} = \frac{U_{JS}}{U_{IS}} K_{IJ} \ (\tau \to 1) \qquad \text{or} \qquad k_{IJ}^{Pot} = \frac{(U_J + U_S) k_J}{(U_I + U_S) k_I} \ (\tau \to 0) \ (1.28)$$

Therefore, while the selectivity coefficient for membranes with completely dissociated or strongly associated and immobile sites depends only on the properties of the solvent, that for membranes with strongly associated and highly mobile sites depends both on the ion-exchange site properties and on the properties of the solvent.

1.1.1.3 LIQUID MEMBRANES WITH ELECTRONEUTRAL ION-CARRIERS

These membranes consist of a suitable inert organic solvent with a dissolved macrocyclic substance which may be an antibiotic (e.g. valinomycin), a macrotetrolide or a cyclic polyether, or with a macromolecular acyclic compound. The structure of these substances forms a polar cavity in which certain ions, chiefly those of the alkali metals, can be enclosed because of ion–dipole interactions, with formation of charged mobile complexes, soluble in organic solvents. This process exhibits a high degree of selectivity and is important in biology, where it is the means of transport of univalent ions across cell membranes, and now also in chemistry, where it gives rise to membrane potentials, selective for some ions. The field has recently been extensively reviewed by Koryta [25].

The theory of these membrane potentials has been given by Eisenman and co-workers [48, 54—57]. This theory is based on the concept of a very thin membrane, so that no assumptions are made concerning the concentration and electric potential distribution, and deviations from the electroneutrality condition are not excluded. Then for a membrane containing ion-carrier S, separating solutions 1 and 2 containing ions I^+ and J^+, and assuming that the formation of complexes IS^+ and JS^+ in the aqueous phases is negligible, the membrane potential is

$$\Delta\varphi_M = \frac{RT}{F}\ln\frac{a_I(1) + \left[\dfrac{U_{JS+}}{U_{IS+}}\dfrac{k_{JS+}}{k_{IS+}}\dfrac{K_{JS+}}{K_{IS+}}\right]a_J(1)}{a_I(2) + \left[\dfrac{U_{JS+}}{U_{IS+}}\dfrac{k_{JS+}}{k_{IS+}}\dfrac{K_{JS+}}{K_{IS+}}\right]a_J(2)} \tag{1.29}$$

where the k values are the distribution coefficients of the complexes between the aqueous and the organic phase and the K values are the stability constants of the complexes in the aqueous phase. The selectivity coefficient is then given by

$$k_{IJ}^{Pot} = \frac{U_{JS+}k_{JS+}K_{JS+}}{U_{IS+}k_{IS+}K_{IS+}} \tag{1.30}$$

As the mobilities of the complexes in the membrane are usually similar, the expression for k_{IJ}^{Pot} reduces to

$$k_{IJ}^{Pot} = \frac{k_{JS+}K_{JS+}}{k_{IS+}K_{IS+}} \tag{1.31}$$

1.1.2 Potentials of Working Ion-Selective Electrodes

It follows from the sections above that the potential of an ISE measured against a reference electrode can be expressed by the general equation

$$E = E_0 \pm \frac{RT}{z_1 F}\ln\left[a_1 + \sum_{i=2}^{n} k_{1i}^{Pot}(a_i)^{z_1/z_i}\right] \tag{1.32}$$

for n ions present in the test solution, the $+$ sign holding for cations and the $-$ sign for anions. The liquid-junction potential(s) and the terms concerning an electrode internal solution with a constant composition are included in the value of E_0. Consequently, the "standard potential" of the electrode, E_0, will be different for each electrode of a certain type and for any set of experimental conditions and will have to be determined experimentally. A typical E vs. pa_X plot for an ISE is depicted in Fig. 1.1. The curvature on plot 1 is caused, for example, by the solubility of the membrane or by the adsorption of substances on the membrane surface, and determines the detection limit attainable with the given electrode. In the presence of an interfering ion the curvature occurs at higher activities of the primary ion and with decreasing activity of the primary ion the slope of the plot decreases until a constant potential, corresponding to the given interferent activity, is exhibited. The three regions, a, b, c, correspond to the situations when the second term in the square brackets in Eq. (1.32) can be neglected with respect to a_1 (the Nernstian-slope region, region *a*), when the two terms are comparable (region *b*) and when a_1 can be neglected with respect to the second term (region *c*).

The theoretical k^{Pot} values, given by the equations above for various electrode types, can serve as a useful, approximate, *a priori* assessment of electrode performance under certain conditions, but good agreement with the experimental values is obtained rather rarely (for a more detailed discussion see Section 1.2). This is due to several simplifications involved in the derivation of the theoretical equations.

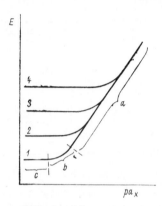

Fig. 1.1 A typical E vs. pa_X dependence for an ion-selective electrode. 1 — primary ion alone, 2, 3, 4 — gradually increasing activity of an interfering species.

First, it is assumed that at least the system in the close vicinity of the membrane–solution interface is in thermodynamic equilibrium, which need not necessarily be true. Second, constant activity coefficients and mobilities of ions in the membrane are assumed, although both values are affected by the presence of other charged or polar species in the system. Third, the membrane is considered to be a homogeneous body with a stationary diffusion potential profile across its whole width; this can be attained only with extremely thin membranes and, in fact, Eq. (1.29) was derived exclusively for such thin membranes. Fourth, only ion-exchange or complexation processes and ion mobilities in the membrane are considered, while in practice, selective adsorption or redox reactions etc. can also contribute to the electrode potential formation.

The theoretical equations above assume the presence of two interface potential differences and a diffusion potential across the membrane. However, with some solid membranes no diffusion potential will be formed, namely, when the membrane is sensitive to a certain ion and another species carries the charge inside the membrane. The insoluble silver halide membrane can serve as an example; it responds to the activity of halide ions in the solution but silver ions act as charge carriers inside the membrane. It has been shown [58, 59] that, in the presence of a single type of anion X^- in solutions 1 and 2 separated by the membrane, the membrane potential is identical with the potential difference measured in a concentration cell containing two electrodes of the second kind, i.e.

$$\Delta \varphi_M = \frac{RT}{F} \ln \frac{a_{X^-}(2)}{a_{X^-}(1)} \tag{1.33}$$

In the presence of two ions, X^- and Y^-, forming insoluble salts with the cation (e.g. chloride and iodide) the electrode potential will be given by Eq. (1.32) for $n = 2$, with a selectivity coefficient approximately given by the ratio of the corresponding solubility products,

$$k_{XY}^{Pot} = \frac{K_S(AgX)}{K_S(AgY)} \tag{1.34}$$

Therefore, in the presence of an ion forming a less soluble precipitate with an ion of the membrane material, the membrane is gradually covered with this precipitate and the electrode becomes selective to the ion. On the other hand, in the presence of ions forming a soluble complex with an ion of the membrane material, with a stability constant sufficiently high for the membrane to dissolve, the electrode will respond to the activity of the complexing agent — an example is the use of silver halide electrodes for the determination of cyanide. The selectivity coefficient is then given essentially by a certain power of the ratio of the diffusion coefficients of cyanide and the halide [60–64]. Another example of materials in which the contribution of the diffusion potential is negligible is chalcogenide electrodes sensitive to heavy metal ions, which exhibit predominantly electronic conductivity. This type of material was thoroughly examined theoretically by Sato [65], who has shown that the electric potentials of these materials depend not only on the activities of the ions in the solution, but also on the activities of the component elements in the solid phase. The potential of substance M_iX_j, dissociating to give M^{j+} and X^{i-} ions in aqueous solutions, is then described by the general formula

$$E_{M_iX_j} = E_{M_iX_j}^0 + \frac{RT}{2ijF} \ln \left[\frac{(a_M^{j+})_i^i (a_X)_s^j}{(a_X^{i-})_i^j (a_M)_s^i} \right] \tag{1.35}$$

where

$$E_{M_iX_j}^0 = \frac{1}{2} \left(E_{M,M^{j+}}^0 + E_{X^{i-},X}^0 \right) \tag{1.36}$$

and subscripts s and l refer to the solid and liquid phase, respectively. The activities in the solid phase then depend on the method of electrode preparation [18, 66, 67] and may vary owing to the effects of the passage of electric current [65], of chemical interactions on contact with solutions [66, 68] and of light [68, 69].

Generally, working thick-solid membranes can be visualized as a bulk phase, through which charge is transported either by ions, usually by means of some defect mechanism (Frenkel or Schottky), or by ions and electrons, neither of these mechanisms being affected by interface phenomena, and very thin surface layers in which interactions with the solution components occur. When in contact with solutions, certain materials form solvated gel layers on their surface and the interface reactions then take place in these. This is notably true of glass electrodes, where the existence of a hydrated film has long been known and for which the film properties have been thoroughly studied (see e.g. [70−74]). A hydrophilic film has also been detected on LaF_3 single-crystal electrodes and its properties have been studied [75, 76]. The exchange reactions taking place at the interface and caused by electrostatic and/or chemical interactions then depend on the surface properties of the solid phase and the kinds of ions in solution. Two electrical double-layers are formed, one in solution, consisting of ions and oriented polar molecules, and the the other in the solid, formed by charged lattice defects [77, 78] which also take part in the charge transport across the interface [79, 80]. These lattice defects, the concentration of which is higher in the surface layer than in the bulk of the solid, are also responsible for the original electric charge of the solid surface, before interaction with the solution [81−83]. On the other hand, chemical interactions between the solid surface and solutions can cause even purely redox sensitive materials to exhibit a certain sensitivity to ions. For example, surfaces of various forms of carbon contain ion-exchange sites and on chemical or electrochemical oxidation in solution are gradually converted into graphite oxide, which exhibits ion-exchange properties and no redox sensitivity [84, 85].

There have been attempts to modify the theoretical equations, in view of the above-mentioned properties of electrode materials, to give a better fit to the experimental results. Two opposite approaches have been taken and applied to glass electrodes. Buck [86, 87] has developed a solid-state theory, involving the Nikolskii ion-exchange and Eisenman n-type non-ideal behaviour concepts, introducing the assumption that only a small fraction of the cations in the glass membrane will be mobile and contribute to the diffusion potential. Consequently, the ion-exchange selectivity is found to be the predominant factor determining the electrode performance. On the other hand, Tadros and Lyklema [88 – 90] conclude from double-layer investigations that ion mobilities in the solvated gel layer on the glass electrode are decisive for the electrode selectivity. Both approaches have their merits and show two separate aspects of a complex problem. Further progress in the theory of solid membranes will probably depend upon progress in solid-state physics.

Several other important aspects of ISE potentials have been recently examined. Metal/salt interface potentials, which are important in electrodes without an internal solution [91], and mixed-sulphide electrode "standard potentials" [92] were studied and it was concluded that the interface potentials are primarily determined by the elemental defect activities in the membrane, and the mixed sulphide electrode standard potentials by the activities of the metal (or sulphur) in the membrane.

Simon and co-workers derived a general equation for the stationary EMF of a cell with an ISE [93], using assumptions similar to those employed in the classical treatments and applied it to silver halide electrodes to assess the selectivity and the detection limit [94]. It was recognized earlier that the detection limit of an ISE is determined by the solubility product of the membrane and an equation was derived for calculation of the limiting potential value [95]. Simon et al. [94] state that the detection limit is determined either by the solubility of the membrane or the silver defect activity in the surface, depending on which is the larger. However, in another paper it is concluded that silver defects do not play a significant role in determining the detection limit of the Ag_2S electrode [96, 112 c].

Finally, the dynamic response of ISE's, important especially in continuous monitoring, has been examined in several papers [97 – 100]. The time-dependence of the measured activity is exponential for ion-exchange membranes [97 – 100], while a square-root dependence is obtained for membranes with electroneutral ion-carriers [100], the diffusion within the boundary layer of the sample solution being the determining factor. The response rate is chiefly affected by the stirring rate and by the direction of sample-activity variation and with neutral ion-carrier membranes also by the extraction capacity [100].

Some theoretical work has also recently been done on liquid membranes; Freiser et al. [101] stressed the importance of the parameters describing solvent extraction in characterizing the electrode response. Carmack and Freiser [102] studied the properties of polymeric-matrix liquid-membrane electrodes and their results indicate ionic charge conduction in the membrane.

The original Eisenman theory for membranes with electroneutral ion-carriers has been criticized on the grounds that the use of the Nernst–Planck equations is questionable with extremely thin membranes [103 – 106]. These authors developed a theory based on the Eyring activation model of ion transport across the membrane. However, this theory is also applicable only to extremely thin membranes. Another group of authors [107] assumes the existence of fixed negative sites in neutral ion-carrier membranes; then the electroneutrality condition can be valid.

An important deviation from the theoretical expectations is encountered with neutral ion-carrier electrodes in the presence of certain anions which are soluble in the membrane [108 – 110]. With increasing salt activity the sensitivity to cations decreases and finally the electrode may respond to anions, with theoretical slope. This phenomenon was studied by Boles and Buck [111] who proposed a generalization of the original Eisenman theory by including considerations of anion solubility, mobility and ion-pair formation. Simon *et al.* [112] have shown that this anion interference can be suppressed by incorporating a suitable lipophilic anion, e.g. tetraphenylborate, into the membrane, thus shifting the anion extraction equilibrium. A decrease in the membrane polarity, e.g. when a suitable polymeric matrix is used, also decreases the anion interference. Most recently, further theoretical work was carried out on various types of ISE (see e.g. refs. [112a, b, c, d, e]).

1.2 IMPORTANCE OF SELECTIVITY COEFFICIENTS AND THEIR DETERMINATION

It is clear from the discussion above that the selectivity coefficient is the basic source of information on the interferences in the ISE response. It is also clear from the relations for the selectivity coefficients of various electrode types that its value will depend upon the experimental conditions, predominantly on the activities of the primary ion and of interfering ions and on the ionic strength. For this reason we consider it preferable to call this quantity the selectivity coefficient (or parameter) rather than the selectivity constant. As has been pointed out, the calculated values of selectivity coefficients give only a rough assessment of the electrode selectivity and more reliable values must be determined experimentally for the given conditions.

There are several experimental methods for the determination of selectivity coefficients, which will be surveyed below. Even these methods give only approximate values of k^{Pot}, as they assume that the electrode behaviour obeys the theory, especially that the electrode response is Nernstian, which need not always be true. Furthermore, the k^{Pot} value obtained holds only for the given experimental conditions and will be different for different ion activities and different ionic strengths. For this reason the electrode selectivity would be better described by a series of k^{Pot} values in a certain ion activity range or by a graph of k^{Pot} vs. activity. As the numerical values determined depend to a certain degree on the method employed, this should always be specified. It has been pointed out [113] that these requirements are unfortunately often not met in the literature. An additional difficulty in practical measurements is that several interfering ions are often present simultaneously; numerical data are then very hard to obtain. Finally, it should be noted that sometimes the reciprocals of the k^{Pot} values defined by Eq. (1.32) are quoted; a decreasing k^{Pot} value then indicates an increasing degree of interference.

The experimental methods for the determination of selectivity coefficients can be classified into two groups, namely, those involving the measurement of the electrode response in separate solutions, one containing the primary ion and the other an interfering ion, and those measuring the electrode potential in solutions containing a mixture of the two ions. The first approach has been criticized (see e.g. [114, 115]), because it does not correspond to the real situation in analyses. This criticism is certainly justified; nevertheless, the values determined by the separate-solution methods often agree reasonably well with those obtained by the mixed-solution methods (see e.g. [114]).

The separate-solution methods follow directly from Eq. (1.32). *Method IA*. If the electrode potentials are measured in solutions with the same activities of the primary (I) and an interfering (J) ion with the same charge, z, [116] then the k_{IJ}^{Pot} value can be calculated from the potential difference

$$(E_J - E_I)/[RT)/(zF)] = \ln k_{IJ}^{Pot} \tag{1.37}$$

Method IB. On the other hand, when a series of measurements is taken in separate solutions of I and J of various activities and the activities for $E_I = E_J$ are found [117, 118], then

$$k_{IJ}^{Pot} = \frac{a_I}{a_J} \tag{1.38}$$

In addition to the criticism mentioned above, it has also been pointed out [96, 119] that, during measurement in solutions of J, an undefined concentration of ion I is formed at the electrode because of an ion-exchange reaction between I and J on the electrode surface, thus introducing an error in the determination of k_{IJ}^{Pot}.

Method IIA. In the first mixed-solution method [96, 119], the electrode potential is measured in a series of solutions containing various activities of the primary ion and a constant activity of the interfering ion. In the E vs. pa_X plot (Fig. 1.2), the "critical" activity a_I is found at the intercept of the extrapolated linear branches; activity a_I and the known activity a_J then determine the k_{IJ}^{Pot} value according to Eq. (1.38). As the horizontal part of the experimental plot in Fig. 1.2 is often subject to irreproducibility and drift, it is better to determine the a_I value as the point at which the difference between the experimental curve and the extrapolated Nernstian dependence equals $18/z$ mV [120].

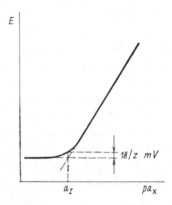

Fig. 1.2 Determination of k_{IJ}^{Pot} in mixed solutions with constant activity of the interfering ion.

Method IIB. The second mixed-solution method is actually the reverse of the first [121]. The primary ion activity, $a_{I^{z+}}$, is maintained constant and that of the interferent is varied (Fig. 1.3); this method has been chiefly used for dealing with H^+-ion interference. The k_{IH}^{Pot} value is then given by

$$k_{IH}^{Pot} = \frac{a_{I^{z+}}}{(a_{H+})^z} \qquad (1.39)$$

These two methods have an advantage in that an assessment of the selectivity can be made directly from the E vs. pa_X plot.

Method IIC. The third mixed-solution method [122] is based on the measurement of the electrode potential in a solution of pure primary ion with activity $a_I (E_I)$, and in a series of solutions with certain activities of the primary ion, a_I', and the interfering ion J, $a_J' (E')$. Combination of the appropriate forms of Eq. (1.32), i.e.

$$E = E^0 \pm \frac{RT}{F} \ln a_I \tag{1.40}$$

$$E' = E^0 \pm \frac{RT}{F} \ln \left[a'_I + k_{IJ}^{\text{Pot}} a'_J \right]$$

gives

$$\left\{ \exp \left[(E - E') F/RT \right] \right\}^{-1} \cdot a_I - a'_I = k_{IJ}^{\text{Pot}} a'_J \tag{1.41}$$

for cations and

$$\left\{ \exp \left[(E - E') F/RT \right] \right\} \cdot a_I - a'_I = k_{IJ}^{\text{Pot}} a'_J \tag{1.42}$$

Fig. 1.3 Determination of k_{IJ}^{Pot} in mixed solutions with constant activity of the primary ion.

for anions. The plot of the left-hand side of Eq. (1.41) or (1.42) against a'_J is a straight line with a slope equal to k_{IJ}^{Pot}. Equations (1.41) and (1.42) can be used for large values of k_{IJ}^{Pot}. With small k_{IJ}^{Pot} values the changes in the electrode potential with variations in the interferent activity are too small; then the procedure must be reversed and the primary ion added to a high activity of the interfering ion. Equations (1.40) change to

$$E = E^0 \pm \frac{RT}{F} \ln \left[a_I + k_{IJ}^{\text{Pot}} a_J \right] \tag{1.43}$$

$$E' = E^0 \pm \frac{RT}{F} \ln \left[a'_I + k_{IJ}^{\text{Pot}} a'_J \right]$$

and their combination yields

$$\left\{ \exp \left[(E - E') F/RT \right] \right\} a'_I - a_I$$
$$= k_{IJ}^{\text{Pot}} \left\{ a_J - \left\{ \exp \left[(E - E') F/RT \right\} a'_J \right\} \right. \tag{1.44}$$

for cations and

$$a_I' - \{\exp\left[(E - E')\, F/RT\right]\}\, a_I$$
$$= k_{IJ}^{Pot} \{\{\exp\left[(E - E')\, F/RT\right]\}\, a_J - a_J'\} \tag{1.45}$$

for anions. The left-hand side of Eq. (1.44) or (1.45) is then plotted against the term in braces on the right-hand side and k_{IJ}^{Pot} is again obtained as the slope of the linear dependence. This method is certainly the least empirical of all, but a disadvantage lies in that it requires *a priori* assessment of the selectivity (large or small k_{IJ}^{Pot}), and is rather cumbersome,

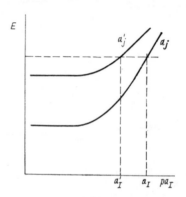

Fig. 1.4 Determination of k_{IJ}^{Pot} according to ref. [123].

and that the exponential terms represent small differences between two large numbers, which is inconvenient from the point of view of accuracy and precision of the determination. For this reason the method has recently been simplified [123], avoiding the use of exponentials (*method IID*). If two solutions are prepared, one containing the primary and the interfering ion at activities a_I and a_J, respectively, and the other at activities a_I' and a_J', then if $a_I > a_I'$ and $a_J < a_J'$, and the electrode potentials measured in these solutions are equal, $E = E'$, then combination of the appropriate forms of Eq. (1.32) gives

$$k_{IJ}^{Pot} = \frac{a_I' - a_I}{a_J - a_J'} \tag{1.46}$$

Therefore, if two series of solutions with various activities of a_I are prepared, one containing interferent activity a_J and the other a_J', and the E vs. a_I dependences are plotted, then k_{IJ}^{Pot} can be calculated for any combination of a_I, a_I', a_J and a_J', for which $E = E'$ (see Fig. 1.4).

The most important conditions for the determination of k^{Pot} by mixed-solution methods have recently been summarized [124] as follows:

(1) only stabilized potentials should be measured; (2) the ionic strength should be substantially lower than 1.0, preferably around 0.1; (3) the concentrations of interferents should be sufficiently high to make a substantial contribution to the electrode potential (e.g. $> 0.1M$ for $k^{Pot} < 0.5$); (4) the concentration of the primary ion should be varied over a wide interval, the initial value being located in the region of the greatest curvature of the E vs. pa_X curve.

1.3 CALIBRATION OF ION-SELECTIVE ELECTRODES

As ISE's respond to activities, knowledge of individual activity coefficients is required if the activity and concentration data are to be correlated. Unfortunately, individual activity coefficients cannot be determined accurately and therefore approximate methods must be used for electrode calibration.

If a sufficient excess of an indifferent electrolyte is present in the solution, then it can be assumed that the ion activity coefficients and the liquid-junction potential(s) depend solely on the indifferent electrolyte concentration and can therefore be included in the value of the "standard" electrode potential, the value of which can be determined for the given electrode by calibration with a solution with a defined concentration of the primary ion. If the solution does not contain sufficient indifferent electrolyte, it is necessary to use a conventional activity scale; this approach was introduced by Bates and Guggenheim [125] for pH-measurements and later applied by Bates and Alfenaar [126] to solutions used in measurements with ISE's. The procedure is based on the convention that the individual molal activity coefficient for chloride ions is calculated on the basis of Guggenheim's concepts [127] from the following form of the Debye−Hückel equation

$$-\log \gamma_{Cl^-} = AI^{\frac{1}{2}}/(1 + 1.5I^{\frac{1}{2}}) \tag{1.47}$$

where I is the molal ionic strength and A equals $0.5108 \ m^{-\frac{1}{2}}$. Since for a salt A_aB_b

$$\left[(a_{A^{b+}})^a (a_{B^{a-}})^b\right]^{1/(a+b)} = a_{\pm(A_aB_b)} \tag{1.48}$$

individual cation activity coefficients can be readily calculated for various chlorides from the γ_{Cl^-} value and from the known mean activities. Then the individual activity coefficients for other anions can be calculated with the help of these cationic activity coefficients.

However, as this procedure is based on the Debye–Hückel limiting law, it fails at higher ionic strengths (> 0.1). For high ionic strengths, a procedure has been developed [128] based on the Robinson–Stokes equation [129], which expresses the mean molal activity coefficient in terms of the ionic strength, the water activity of the solution and the hydration number. It can be assumed that chloride ions are virtually unhydrated and then the individual activity coefficients can readily be calculated for solutions of chlorides from the equations

$$\log \gamma_+ = \log \gamma_\pm + 0.00782(h_+ - h_-)\, m\Phi$$
$$\log \gamma_- = \log \gamma_\pm + 0.00782(h_- - h_+)\, m\Phi \tag{1.49}$$

where h_+ and h_- are the hydration numbers of the cation and anion, respectively, m is the solution molality and Φ is the electrolyte osmotic coefficient. This procedure is valid up to concentrations of $6m$.

By use of solutions with individual activities calculated in this way the activity of an ion in a test solution, $pa(X)$, can be determined by comparing the electrode potentials measured in the test solution and a standard solution of activity $pa(S)$:

$$pa(X) = pa(S) + \frac{[E(X) - E(S)]\, zF}{2.303RT} \tag{1.50}$$

Equation (1.50) holds when the liquid-junction potentials in solutions X and S can be considered identical; otherwise a correction must be introduced, e.g. by using the Henderson equation, (1.7).

When an electrode is to be calibrated in the region of very low activities, difficulties arise from poorly defined changes in the activities, caused by adsorption, hydrolytic processes etc. Then it is possible to use metal buffers (see e.g. [130]), based on complexation reactions. When a stable 1 : 1 metal complex is formed, ML, then the activity of the metal can be determined from the equation

$$pM = \log K_{ML} + \log \frac{c_L}{c_{ML}} - \log \alpha \tag{1.51}$$

where K_{ML} is the stability constant, c_L and c_{ML} are the overall concentrations of the ligand and the complex, respectively, and α is the Ringbom side-reaction function [131]. By varying the ligand concentration and the pH, the metal ion activity can be adjusted to values down to ca. $10^{-20}M$. Ethylenediaminetetra-acetic acid and its analogues are well suited for this purpose. (For practical problems of electrode calibration see Chapter 3.)

1.4 ELECTRODES WITH ADDITIONAL MEMBRANES

ISE's can also be employed for the determination of various non-ionic substances, by using additional membranes in combination with a suitable chemical equilibrium selective for the given species, in which an ion sensed by a conventional ISE is formed. Gas-sensing and enzyme electrodes belong to this group. Some electrodes with additional membranes operate amperometrically.

1.4.1 Gas-Sensing Electrodes

These electrodes contain a membrane permeable to the required gas and separating the test solution from an internal solution, in which an ISE and a reference electrode are immersed. The internal solution interacts with the gas to form the ion sensed by the ISE. The first electrode of this kind was the carbon dioxide electrode [132, 133], where the gas permeates through a membrane into a sodium bicarbonate medium, changing its pH, and these changes are monitored by a pH-sensitive glass electrode. Later, electrodes for other gases, e.g. NH_3, SO_2, NO_2, HCN, etc., appeared (see Chapter 2 and the appropriate sections of Chapter 4). A brief and so far the only general treatment of these electrodes is that given by Ross *et al.* [134]. These authors assume steady-state transport of the gas through the membrane and establishment of equilibrium of the chemical reaction with the internal electrolyte. For the reaction

$$a\text{A} + b\text{B} \;\rightleftharpoons\; i\text{I} + p\text{P} \tag{1.52}$$

where A is the diffusing gas, B is the component of the internal solution that reacts with A, I is the ion formed that is sensed by the ISE and P is the sum of the other products, the law of mass action gives

$$[\text{I}] = \left\{ \frac{k_{eq} \cdot [\text{B}]^b}{[\text{P}]^p} \right\}^{(1/i)} \cdot [\text{A}]^{(a/i)} \tag{1.53}$$

Generally it can be assumed that [B] and [P] are high compared with [A] and [I] and, the membrane being impermeable to B and P, are virtually constant; then

$$[\text{I}] = \text{const.} \, [\text{A}]^{(a/i)} \tag{1.54}$$

and consequently the concentration of I sensed by the ISE will be proportional to the concentration of gas A in the test solution. The electrode will respond to gas A with a slope of $2.3(RT/nF)\,(a/i)$. In order to satisfy the condition that [B] and [P] should be constant, the equilibrium constant, K_{eq}, must not be too high; on the other hand, too low a K_{eq} value would cause the activity of I to be so low that it would not be detected by the ISE. The electrode sensitivity also depends on the concentration of the internal solution, which must be selected on the basis of the sample concentration (see Fig. 2.5). Finally, various side-reactions of the test substance must be considered, as well as the possibility of interference from water transported through the membrane. In order to minimize the latter effect, the osmotic pressures and temperatures of the test and internal solutions must be maintained at similar values. The performance of a gas-sensing electrode will greatly depend upon the rate at which the equilibrium is established, which primarily depends upon the rate of gas transport through the membrane and the rate of the chemical reaction. Assuming that a steady state is established rapidly after a sharp change in concentration of the gas in the test solution and that a linear concentration gradient is formed across the membrane, Fick's law can be employed to describe the rate at which equilibrium is established [134]. If the diffusing gas undergoes side-reactions in the internal electrolyte, then its total concentration in the internal electrolyte is

$$c_{tot} = c + c_{side} \qquad (1.55)$$

where c is the concentration of the neutral species. Assuming that dc_{side}/dc is very small or at least constant, the time required to approach equilibrium is given by

$$t = \frac{d_1 \cdot d_s}{D \cdot k}\left[1 + \frac{dc_{side}}{dc}\right]\ln\frac{\Delta c}{\varepsilon c_{sample}} \qquad (1.56)$$

where d_1 is the thickness of the film of internal electrolyte, d_s is the thickness of the membrane, D is the diffusion coefficient, k is the distribution coefficient of the gas between the solutions and the membrane, $\Delta c = c_{sample} - c_{init}$ is the magnitude of the concentration change in the test solution and ε is a measure of the approach to equilibrium, defined by.

$$\varepsilon = \left|\frac{c_{sample} - c}{c_{sample}}\right| \qquad (1.57)$$

From Eq. (1.56) it follows that, for rapid response, it is necessary that d_1 and d_s be as low as possible and D and k as high as possible, i.e. the membrane and the internal electrolyte film should be very thin and the membrane should permit very rapid transport of the gas. Further it can be seen that the time response will be practically independent of the magnitude of Δc if the change is from low to high concentration ($\Delta c \approx c_{sample}$), while for the reverse change the response time will increase with increasing Δc ($\Delta c \approx c_{init}$). It is also clear that slow transport of the gas through the membrane will render the electrode poorly sensitive and will also practically determine the detection limit.

A graphical method has been developed for evaluating the dynamic measuring range (i.e. the range over which the measured signal is a linear function of the test substance activity) for gas-sensing electrodes [135].

1.4.2 Enzyme Electrodes

Enzyme-catalysed reactions are marked by a high selectivity for certain substrates which are mostly difficult to determine. As ions are usually formed during these reactions, it is possible to determine the substrate (or an enzyme inhibitor, activator or the enzyme itself) by monitoring the activity of the ion. This is the basis of enzyme electrodes. The enzyme can be added to the test solution, the enzymic reaction can take place in a reactor before the measuring cell, or the enzyme can be immobilized on the surface of an ISE in some polymeric matrix; the last type has been developed chiefly by Guilbault and co-workers [17]. This type of electrode is certainly promising, especially in the fields of biochemistry, biology and medicine, but so far its use is rather empirical. As examples, the following systems can be given.

Amygdalin electrode [136]:

(CN⁻-sensitive ISE)

L-Amino-acid electrode [137, 138]:

$$R-CHNH_2COO^- + O_2 + H_3O^+ \xrightarrow{\text{L-amino-acid oxidase}}$$
$$R-COCOO^- + NH_4^+ + H_2O_2$$
$$(NH_4^+\text{-sensitive ISE})$$

Urea electrode [139 – 141]:

$$CO(NH_2)_2 + 2 H_3O^+ \xrightarrow{\text{urease}} CO_2 + 2 NH_4^+ + 2 H_2O$$
$$(NH_4^+\text{-sensitive ISE})$$

1.4.3 Electrodes Differentiating Organic Substances with Different Structures

There are also electrodes capable of differentiating enantiomers. A chiral, electrically neutral carrier contained in an ISE forms a complex with α-phenylethylammonium ion; one of the enantiomers causes a greater change in the electrode potential than the other [142, 143].

The selectivity of the "immunoelectrode" [144] is based on immunological reactions.

1.5 ANALYTICAL APPLICATION OF ION-SELECTIVE ELECTRODES

It follows from the discussion above that ISE's can be used for analysis of a great variety of materials, depending on the selectivity coefficients. The detection limit is essentially determined by the membrane solubility, i.e. it mostly varies around a concentration of 10^{-4}–$10^{-6}M$; however, substantially lower activities can sometimes be measured if metal buffers are used. In this possibility lies a great advantage for theoretical studies of solution equilibria.

The determination error, not considering sources of error other than those concerned with the electrode system, is chiefly due to changes in the electrode potential with temperature and to potential drift. Both these differ for various electrodes and there are not enough theoretical and experimental data for their prediction. In continuous monitoring the rate of the electrode response may contribute to the overall error – the dependence of E on time was briefly discussed in Section 1.1.2. Mostly there are no problems with solid-membrane electrodes, where the response time varies from units to hundreds of milliseconds, but with liquid-membrane electrodes it may be as much as several minutes. The response time is very important with gas-sensing electrodes, where it actually determines the detection limit.

Generally it can be said that potentiometry with ISE's has established itself as an excellent method for the rapid, simple and cheap routine analysis of inorganic materials, in biology, medicine, pollution control, the food industry, etc., and is especially well suited for continuous measurements in flow-through systems. It seems that the stage of rapid, extensive and frequently rather empirical development of the field is now over and further progress will predominantly depend on improvements in the theory.

REFERENCES

1. M. Cremer: *Z. Biol.*, **47**, 562 (1906).
2. F. Haber and Z. Klemensiewicz: *Z. Physik. Chem.*, **67**, 385 (1909).
3. B. Lengyel and E. Blum: *Trans. Faraday Soc.*, **30**, 461 (1934).
4. O. Tomíček and R. Půlpán: *Chem. Listy*, **49**, 497 (1955); *Collection Czech. Chem. Commun.*, **21**, 1444 (1956).
5. G. Eisenman (ed.): *Glass Electrodes for Hydrogen and Other Cations, Principles and Practice*, Dekker, New York (1967).
6. F. Haber: *Ann. Physik*, **26**, 927 (1908).
7. H. J. C. Tendeloo: *J. Biol. Chem.*, **113**, 333 (1936).
8. H. J. C. Tendeloo and A. Krips: *Rec. Trav. Chim.*, **76**, 703 (1957).
9. H. J. C. Tendeloo and A. Krips: *Rec. Trav. Chim.*, **78**, 177 (1959).
10. R. B. Fischer and R. F. Babcock: *Anal. Chem.*, **30**, 1732 (1958).
11. J. S. Parsons: *Anal. Chem.*, **30**, 1262 (1958).
12. E. Pungor and E. Hollós-Rokosinyi: *Acta Chim. Acad. Sci. Hung.* **27**, 63 (1961).
13. M. S. Frant and J. W. Ross: *Science*, **154**, 1553 (1966).
14. K. Sollner and G. M. Shean: *J. Am. Chem. Soc.*, **86**, 1901 (1964).
15. J. W. Ross, Jr.: *Science*, **156**, 1378 (1967).
16. Z. Štefanac and W. Simon: *Chimia*, **20**, 436 (1966).
17. G. G. Guilbault: *Pure Appl. Chem.*, **25**, 727 (1971).
18. V. Majer, J. Veselý and K. Štulík: *Anal. Letters*, **6**, 577 (1973).
19. W. Nernst: *Z. Phys. Chem.*, **2**, 613 (1888).
20. W. Nernst: *Z. Phys. Chem.*, **4**, 129 (1889).
21. M. Planck: *Ann. Phys. Chem.*, **39**, 196 (1890).
22. M. Planck: *Ann. Phys. Chem.*, **40**, 561 (1890).
23. P. Henderson: *Z. Physik. Chem.*, **59**, 118 (1907).
24. P. Henderson: *Z. Physik. Chem.*, **63**, 325 (1908).
25. J. Koryta: *Ion-Selective Electrodes*, p. 12. Cambridge University Press, Cambridge (1975).
26. G. N. Lewis and L. W. Sargent: *J. Am. Chem. Soc.*, **31**, 363 (1909).
27. M. Spiro: *Electrochim. Acta*, **11**, 569 (1966); R. G. Picknett: *Trans. Faraday Soc.*, **64**, 1059 (1968); W. H. Smyrl and J. Newman: *J. Phys. Chem.*, **72**, 4660 (1968).
28. F. G. Donnan: *Z. Elektrochem.*, **17**, 572 (1911).
29. K. Sollner: *Z. Elektrochem.*, **36**, 36 (1930).
30. K. Sollner: *J. Macromol. Sci.-Chem.*, A3, 1 (1969).
31. T. Teorell: *Trans. Faraday Soc.*, **33**, 1053 (1937).
32. T. Teorell: *Z. Elektrochem.*, **55**, 460 (1951).
33. T. Teorell: *Prog. Biophys. Biophys. Chem.*, **3**, 300 (1953).
34. T. Teorell: *Disc. Faraday Soc.*, **21**, 9 (1956).
35. T. Teorell: *Biophys. J.*, **2**, Part 2, Suppl., 27 (1962).
36. K. H. Meyer and J. F. Sievers: *Helv. Chim. Acta*, **19**, 649, 665 (1936).
37. K. H. Meyer and J. F. Sievers: *Trans. Faraday Soc.*, **33**, 1073 (1937).
38. B. P. Nikolskii: *Zh. Fiz. Khim.*, **10**, 495 (1937).
39. K. Horowitz: *Z. Phys.*, **15**, 369 (1923).
40. H. Schiller: *Ann. Phys.*, **74**, 105 (1924).
41. B. P. Nikolskii: *Zh. Fiz. Khim.*, **27**, 724 (1953).
42. B. P. Nikolskii and M. M. Shults: *Vestn. Leningrad. Univ.* **4**, 73 (1963).
43. B. P. Nikolskii and M. M. Shults: *Zh. Fiz. Khim.*, **36**, 1327 (1962).

44. B. P. Nikolskii, M. M. Shults and A. A. Beliyustin: *Dokl. Akad. Nauk SSSR* **144**, 844 (1962).
45. G. Karreman and G. Eisenman: *Bull. Math. Biophys.*, **24**, 413 (1962).
46. F. Conti and G. Eisenman: *Biophys. J.*, **5**, 247 (1965).
47. F. Conti and G. Eisenman: *Biophys. J.*, **5**, 511 (1965).
48. G. Eisenman: in R. A. Durst (ed.), *Ion-Selective Electrodes*, Chapter I., NBS Special Publ. 314, Washington (1969).
49. J. P. Sandblom, G. Eisenman and J. L. Walker, Jr.: *J. Phys. Chem.*, **71**, 3862 (1967).
50. J. P. Sandblom, G. Eisenman and J. L. Walker, Jr.: *J. Phys. Chem.*, **71**, 3871 (1967).
51. G. Eisenman: *Anal. Chem.*, **40**, 310 (1968).
52. J. L. Walker, Jr., G. Eisenman and J. P. Sandblom: *J. Phys. Chem.*, **72**, 978 (1968).
53. J. P. Sandblom: *J. Phys. Chem.*, **73**, 249 (1969).
54. S. Ciani, G. Eisenman and G. Szabo: *J. Membrane Biol.*, **1**, 1 (1969).
55. G. Eisenman, S. Ciani and G. Szabo: *J. Membrane Biol.*, **1**, 294 (1969).
56. G. Szabo, G. Eisenman and S. Ciani: *J. Membrane Biol.*, **1**, 346 (1969).
57. G. Szabo, G. Eisenman and S. Ciani: in *Proc. Coral Gables Conf. Physical Principles of Biological Membranes* (1968), Gordon and Breach, New York (1969).
58. R. P. Buck: *Anal. Chem.*, **40**, 1432 (1968).
59. R. P. Buck: *Anal. Chem.*, **40**, 1439 (1968).
60. W. Jaenicke: *Z. Elektrochem.*, **55**, 648 (1951).
61. W. Jaenicke: *Z. Elektrochem.*, **57**, 843 (1953).
62. W. Jaenicke and M. Haase: *Z. Elektrochem.*, **63**, 521 (1959).
63. D. H. Evans: *Anal. Chem.*, **44**, 875 (1972).
64. G. P. Bound, B. Fleet, H. von Storp and D. H. Evans: *Anal. Chem.*, **45**, 788 (1973).
65. M. Sato: *Electrochim. Acta*, **11**, 361 (1966).
66. J. Veselý, O. J. Jensen and B. Nicolaisen: *Anal. Chim. Acta*, **62**, 1 (1972).
67. J. Růžička and E. H. Hansen: *Anal. Chim. Acta*, **63**, 115 (1973).
68. J. Veselý: *Collection Czech. Chem. Commun.*, **39**, 710 (1974).
69. A. Marton and E. Pungor: *Anal. Chim. Acta*, **54**, 209 (1971).
70. B. Karlberg: *J. Electroanal. Chem.*, **45**, 127 (1973).
71. B. Karlberg: *J. Electroanal. Chem.*, **49**, 1 (1974).
72. A. Wikby: *Electrochim. Acta*, **19**, 329 (1974).
73. A. Wikby and B. Karlberg: *Electrochim. Acta*, **19**, 323 (1974).
74. T. Matsushima and M. Enyo: *Electrochim. Acta*, **19**, 117, 125, 131 (1974).
75. M. J. D. Brand and G. A. Rechnitz: *Anal. Chem.*, **42**, 478 (1970).
76. J. Veselý and K. Štulík: *Anal. Chim. Acta*, **73**, 157 (1974).
77. E. P. Honig: *Trans. Faraday Soc.*, **65**, 2248 (1969).
78. E. P. Honig: *Nature*, **225**, 537 (1970).
79. T. B. Grimley and N. F. Mott: *Disc. Faraday Soc.*, **1**, 3 (1947).
80. T. B. Grimley: *Proc. Roy. Soc. London*, **A201**, 40 (1950).
81. K. Lehovec: *J. Chem. Phys.*, **21**, 1123 (1953).
82. K. L. Klilever: *J. Phys. Chem. Solids*, **27**, 765 (1966).
83. F. Trantweiler: *Phot. Sci. Eng.*, **12**, 98 (1968).
84. V. Majer, J. Veselý and K. Štulík: *J. Electroanal. Chem.*, **45**, 113 (1973).
85. M. Štulíková and K. Štulík: *Chem. Listy*, **68**, 800 (1974).

86. R. P. Buck: *Anal. Chem.*, **45**, 654 (1973).

87. R. P. Buck, J. H. Boles, R. D. Porter and J. A. Margolis: *Anal. Chem.*, **46**, 255 (1974).

88. T. F. Tadros and J. Lyklema: *J. Electroanal. Chem.*, **17**, 267 (1968).

89. T. F. Tadros and J. Lyklema: *J. Electroanal. Chem.*, **22**, 1 (1969).

90. T. F. Tadros and J. Lyklema: *J. Electroanal. Chem.*, **22**, 9, (1969).

91. R. P. Buck and V. R. Shepard, Jr.: *Anal. Chem.*, **46**, 2097 (1974).

92. M. Koebel: *Anal. Chem.*, **46**, 1559 (1974).

93. H.-R. Wuhrmann, W. E. Morf and W. Simon: *Helv. Chim. Acta*, **56**, 1011 (1973).

94. W. E. Morf, G. Kahr and W. Simon: *Anal. Chem.*, **46**, 1538 (1974).

95. D. J. Crombie, G. J. Moody and J. D. R. Thomas: *Anal. Chim. Acta*, **80**, 1 (1975)

96. E. Pungor and K. Tóth: *Analyst*, **95**, 625 (1970).

97. G. A. Rechnitz and H. F. Hameka: *Z. Anal. Chem.*, **214**, 252 (1965).

98. R. P. Buck: *J. Electroanal. Chem.*, **18**, 363 (1968).

99. K. Tóth and E. Pungor: *Anal. Chim. Acta*, **64**, 417 (1973).

100. W. E. Morf, E. Lindner and W. Simon: *Anal. Chem.*, **47**, 1596 (1975).

101. H. J. James, G. P. Carmack and H. Freiser: *Anal. Chem.*, **44**, 853 (1972).

102. G. D. Carmack and H. Freiser: *Anal. Chem.*, **47**, 2249 (1975).

103. P. Läuger and G. Stark: *Biochim. Biophys. Acta*, **211**, 458 (1970).

104. G. Stark and R. Benz: *J. Membrane Biol.*, **5**, 133 (1971).

105. G. Stark, B. Kettener, R. Benz and P. Läuger: *Biophys. J.*, **11**, 981 (1971).

106. B. Kettener, B. Neumcke and P. Läuger: *J. Membrane Biol.*, **5**, 225 (1971).

107. O. Kedem, M. Perry and R. Bloch: *IUPAC Intern. Symp. Selective Ion-Sensitive Electrodes*, Cardiff, 1973, G. J. Moody (ed.), paper No. 44.

108. S. Lal and G. D. Christian: *Anal. Letters*, **3**, 11 (1970).

109. L. A. R. Pioda, V. Stankova and W. Simon: *Anal. Letters*, **2**, 665 (1969).

110. M. S. Frant and J. W. Ross: *Science*, **167**, 987 (1970).

111. J. H. Boles and R. P. Buck: *Anal. Chem.*, **45**, 2057 (1973).

112. W. E. Morf, D. Ammann and W. Simon: *Chimia*, **28**, 65 (1974).

112a. A. P. Thoma, A. Viviani-Nauer, S. Arvavitis, W. E. Morf and W. Simon: *Anal. Chem.*, **49**, 1567 (1977).

112b. R. P. Buck: *Anal. Chem.*, **48**, 23R (1976).

112c. R. P. Buck: *CRC Crit. Rev. Anal. Chem.*, 323 (1976).

112d. A. Shatkay: *Anal. Chem.*, **48**, 1039 (1976).

112e. E. Pungor, K. Tóth and E. Lindner: *Anal. Chem.*, **48**, 1071 (1976).

113. G. J. Moody and J. D. R. Thomas: *Selective Ion-Sensitive Electrodes*, p. 20, Merrow, Watford (1971).

114. Ref. 113, p. 13.

115. R. P. Buck: *Anal. Chim. Acta*, **73**, 321 (1974).

116. G. Eisenman, D. O. Rudin and J. U. Casby: *Science*, **126**, 831 (1957).

117. G. A. Rechnitz, M. R. Kresz and S. B. Zamochnick: *Anal. Chem.*, **38**, 973 (1966).

118. G. A. Rechnitz and M. R. Kresz: *Anal. Chem.*, **38**, 1786 (1966).

119. E. Pungor and K. Tóth: *Anal. Chim. Acta*, **47**, 291 (1969).

120. *Orion Res. Newsletter*, **1**, No. 5 (1969).

121. Ref. 113, p. 16.

122. K. Srinivasan and G. A. Rechnitz: *Anal. Chem.*, **41**, 1203 (1969).

123. G. Baum and M. Lynn: *Anal. Chim. Acta*, **65**, 393 (1973).
124. A. Hulanicki and Z. Augustowska: *Anal. Chim. Acta*, **78**, 261 (1975).
125. R. G. Bates and E. A. Guggenheim: *Pure Appl. Chem.*, **1**, 163 (1960).
126. R. G. Bates and M. Alfenaar: in R. A. Durst (ed.), *Ion-Selective Electrodes*, Chapter 6, NBS Special Publ. 314, Washington (1969).
127. E. A. Guggenheim: *J. Am. Chem. Soc.*, **52**, 1315 (1930).
128. R. G. Bates, B. R. Staples and R. A. Robinson: *Anal. Chem.*, **42**, 867 (1970).
129. R. H. Stokes and R. A. Robinson: *J. Am. Chem. Soc.*, **70**, 1870 (1948).
130. R. Blum and M. H. Fog: *J. Electroanal. Chem.*, **34**, 485 (1972).
131. A. Ringbom: *Complexation in Analytical Chemistry*, Interscience, New York (1963).
132. R. W. Stow, R. F. Baer and B. F. Randall: *Arch. Phys. Med. Rehabil.*, **38**, 646 (1957).
133. W. Severinghaus and A. F. Bradley: *J. Appl. Physiol.*, **13**, 515 (1958).
134. J. W. Ross, J. H. Riseman and J. A. Krueger: *IUPAC Intern. Symp. Selective Ion-Sensitive Electrodes*, Cardiff 1973, G. J. Moody (ed.), p. 473, Butterworths, London (1973).
135. E. H. Hansen and N. R. Larsen: *Anal. Chim. Acta*, **78**, 459 (1975).
136. R. A. Llenado and G. A. Rechnitz: *Anal. Chem.*, **44**, 468 (1972).
137. G. G. Guilbault and E. Hrabánková: *Anal. Chem.*, **42**, 1779 (1970).
138. G. G. Guilbault and E. Hrabánková: *Anal. Letters*, **3**, 53 (1970).
139. S. A. Katz and G. A. Rechnitz: *Z. Anal. Chem.*, **196**, 248 (1963).
140. G. G. Guilbault and G. Nagy: *Anal. Chem.*, **45**, 417 (1973).
141. G. G. Guilbault and G. Nagy: *Anal. Letters*, **6**, 301 (1973).
142. J. M. Lehn, A. Moradpour and J. P. Behr: *J. Am. Chem. Soc.*, **97**, 2532 (1975).
143. A. P. Thoma, Z. Cimermann, U. Fiedler, D. Bedeković, M. Güggi, P. Jordan, K. May, E. Pretsch, V. Prelog and W. Simon: *Chimia*, **29**, 344 (1975).
144. J. Janata: *J. Am. Chem. Soc.*, **97**, 2914 (1975).

Chapter 2

INSTRUMENTATION

2.1 ION-SELECTIVE ELECTRODE CONSTRUCTION

2.1.1 Introduction

The construction of ISE's depends primarily on the physico-chemical nature of the electroactive substance and on the purpose for which the electrode is intended and there may be differences in the designs of electrodes from various manufacturers. The electrodes made by various companies for the determination of a certain ion usually contain the same electroactive substance, but they often differ even in their basic characteristics, such as the detection limit and selectivity. These differences are due to the effect of small changes in the composition of the electroactive substances (e.g. with chalcogenide membranes) or of the use of different solvents, plasticizers and polymeric matrix supports of liquid ISE's (these will now be denoted by LISE). Large variations in behaviour are often encountered with electrodes of one type made by the same manufacturer, because of wearing out during measurement and handling in general, and aging. For example, miniature holes are formed on the surface of mixed-membrane ISE's, such as $CuS + Ag_2S$ [1]. Undesirable changes in function can also stem from faulty sealing of the membrane, irreversible changes in the internal contact system, etc. These frequent variations in the ISE properties (which are generally more likely to occur with LISE's) necessitate frequent electrode calibration and sometimes repolishing of the surface of solid membranes or replacement of those of LISE's. As ISE construction has undergone constant development since *ca.* 1965, the published data must be compared cautiously with data obtained with a particular electrode. Sometimes different electroactive substances are employed by different manufacturers for the construction of a particular electrode i.e.g. the Orion 94-29 Cu^{2+}-sensitive electrode contains $CuS + Ag_2S$, but the Radiometer F 1112 Cu or Crytur 29-17 electrodes are based on $Cu_{2-x}Se$; the Corning-EEL (47 613200) K^+-sensitive electrode employs substituted potassium phenylborates, while the Philips IS 560-K electrode contains valinomycin).

ISE construction has recently been developing towards greater flexibility, using replaceable membranes or electroactive substances in a single electrode body [2, 3]. Examples are the Orion 93 series of LISE's or Růžička's "Selectrode" [4], manufactured by Radiometer. Other ISE's with replaceable membranes have also been described [5−7]. From the point of view of their construction it is possible to differentiate ISE's into the following groups:

(1) ISE's with an internal reference solution and internal reference electrode;

(2) ISE's with a solid contact on the inner side of the membrane or liquid film (solid-state and liquid-state electrodes);

(3) ISE's with additional membranes (gas-sensing and enzyme electrodes);

(4) microelectrodes;

(5) combined ISE's.

2.1.2 Electrodes with an Internal Reference Solution and Internal Reference Electrode

These ISE's (see Fig. 2.1) represent the classical type of membrane electrode, which is still used for a majority of glass pH and pM electrodes, fluoride ISE's and most LISE's, although types of these ISE's with a solid contact system have also been described (see Section 2.1.3).

The requirements placed on the internal reference system of an ISE primarily involve long-term stability over a wide temperature range, as errors arising from instability of the electrode potential directly contribute to the determination error. Silver chloride or bromide internal reference electrodes are most often used. Long-term stability is achieved more readily with robust electrodes (see e.g. [8]), where the danger of dissolution or peeling of silver halide films because of temperature fluctuations and different dilatation coefficients is not so imminent. Dissolution of the halide from the electrode and its reprecipitation in the internal solution can be prevented by saturation of the internal standard solution with the halide at an elevated temperature and by adding powdered silver halide; the stability of the ISE potential is then also improved. A silver/silver nitrate solution system can also be employed as an internal reference electrode (with Ag_2S-electrodes); the silver nitrate solution must be protected against reduction caused by light or by contact with silicone rubber. With NO_3^--sensitive LISE's, strong radiation effects (above 10^5 rad) can also lead to changes in the internal reference system [9]. The internal reference solution is selected so that the activities of the test ion and of the ion participating in the reference electrode reaction are effectively buffered; these activities should not be affected by prolonged contact with the electrode body and the membrane. Therefore, rather concentrated solutions are mostly used, e.g.:

$Ag|AgCl, 0.1M\ CaCl_2|membrane|$ (Ca^{2+}-sensitive ISE)

$Ag|AgBr, 0.1M\ KF + 0.1M\ KBr|LaF_3|$ (F^--sensitive ISE)

Although very precise measurements with any ISE require thermostatic control and thermal equilibrium in the whole cell, so that standards and

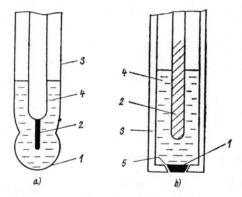

Fig. 2.1 An ISE with internal reference solution and an internal reference electrode. (a) — glass, (b) — solid-state, 1 — membrane, 2 — internal reference electrode, 3 — electrode body, 4 — internal reference solution, 5 — cement.

samples are measured at the same temperature, modern instruments and also a suitable choice of the internal reference system make it possible to attain sufficient precision even when the standards and samples are at a different temperature (isopotential technique, see p. 81). Suitable choice of the internal reference system can lead to a substantial decrease in the cell temperature coefficient (see Section 2.5). This procedure is advantageous when the ISE as a whole has a temperature coefficient of sign opposite to that of the external reference electrode; classical ISE's offer a wider choice in this respect than ISE's with solid contacts.

When measurements are to be made in solutions containing non-aqueous solvents, it may be advantageous to employ the same medium in the internal electrode system, as the response rate and the determination error may depend on the nature of the internal solution [10].

2.1.3 Electrodes with Solid Internal Contacts
(Solid-State and Liquid-State Electrodes)

The construction of classical membrane ISE's is the source of several functional and constructional difficulties. For example, miniaturization and electrical and liquid-proof insulation of the membrane are difficult. Further, when measurements are carried out at higher temperatures or pressures, it is necessary to compensate the pressure difference between the inside and outside of the electrode, to prevent damage to the seal or the membrane.

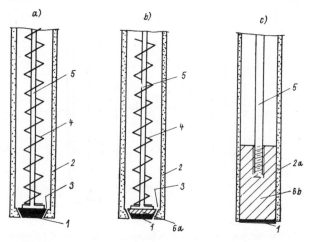

Fig. 2.2 All solid-state ISE. (a) — common type, (b) — ISE with an intermediary layer, c — Růžička's "Selectrode". 1 — membrane or electroactive film, 2 — insulating electrode body, 2a — "Teflon", 3 — cement, 4 — spring, 5 — contact, 6a — intermediary layer, 6b — "Teflon" + graphite mixture.

The solid contact on the internal side of the membrane (see Fig. 2.2) can be formed either directly by an electronic conductor (a metal, graphite), or with some intermediary layer placed between the ion-sensitive membrane and the electronic conductor in order to improve the stability and suppress polarization of the membrane/contact interface. For example, an AgI layer is placed between the Ag_2S membrane and Ag contact in S^{2-}/Ag^+-sensitive electrodes, or a layer of a heavy univalent metal halide (Tl, Cu, Ag) [11–13] is used in glass electrodes.

In ion-sensitive electrodes made of materials other than glass, which can be considered as predecessors of contemporary ISE's, direct contacts were sometimes used, obtained e.g. by fusing or soldering a conductor onto minerals; however, imperfect technique, variability in the composition of the electroactive substances and a lack of knowledge of the dependence of potential stability on the kind of contact [14, 15] complicated the interpretation of measurements. Attempts to elucidate the thermodynamics in such systems have only recently been made [15 – 17]. These constructions were proposed repeatedly [18] (see also [19]) and a silver contact on silver halides and copper contact on $Cu_{2-x}Se$ have proven comparable with classical membrane construction. On the other hand, Au, Al or conductive graphite lacquer contacts are unsatisfactory in all halide electrodes [18]. Conductive epoxy resins containing 80–85 % w/w of a metal (preferably Bi or Pb) have been proposed as contacts for fluoride ISE's [20]. However, these electrodes cannot be compared with ISE's of classical construction, chiefly from the point of view of long-term potential reproducibility. This is perhaps because the LaF_3/metal interface may be polarized by currents as low as those passing through the circuit during measurement with modern direct-reading instruments.

A special case of an electrode with solid contact is Růžička's "Selectrode" [4, 21, 22] (see Fig. 2.2). About 2–4 mg of an electroactive substance is mechanically deposited on a pressed rod of a homogenized mixture of graphite and "Teflon" (pressure of $1 - 2 \times 10^3$ atm, weight ratio ca. 1 : 4). The addition of "Teflon" suppresses the electroactivity of graphite (see e.g. [4, 23]), but the rod is conductive and functions as a contact. The "Selectrode" is especially suitable for electroactive substances which have predominantly electronic conductivity and can be directly applied in powdered form. A solid conductive support with a membrane or film containing a liquid electroactive substance is the basis of liquid-state and coated-wire ISE's. A "Selectrode" (the mixture is pressed at 100 atm [4, 24]) or carbon paste (e.g. [25, 26]) can be used as a support for liquid-state electrodes; other constructions have also been reported [27 – 29]. In coated-wire ISE's [29a] the electroactive substance is dispersed in a polymeric matrix film deposited on a conductor, usually a platinum wire [30, 31]. For example, a Ca^{2+}-ISE is obtained by repeated immersion of a Pt wire (ca. 0.5 mm thick) in a 6 : 1 mixture of 5 % PVC dissolved in cyclohexanol and a Ca^{2+}-selective ion-exchanger, followed by evaporation of the solvent [30]; the ion-exchanger content is ca. 1–25 % w/w. With coated-wire ISE's, changes in the E vs. a relation were observed, e.g. a shorter linear response region for the ClO_4^--ISE [32], and a dependence of the properties of valinomycin/PVC K^+-ISE's on the thickness of the film on the wire [33]. Heterogeneous electrodes with a Cu_2S + silicone rubber mixture deposited on a Pt or Cu support

[34] can also be considered as a form of coated-wire ISE, whereas a mixture of chalcogenides on an Ag support [35] rather resembles classical ion-sensitive electrodes. A similar situation exists with "layer electrodes" [35a]. A common property of liquid-state and coated-wire electrodes is the theoretically undefined potential difference at the support/electroactive substance interface; consequently, the electrode quality may vary and it is so far impossible to predict the electrode behaviour. It has been shown that the PVC-matrix ISE internal reference system is most probably formed by an oxygen electrode [35b].

2.1.4 Solid Membrane Electrodes

In these ISE's the membranes are sealed or otherwise fixed in the end of a body made of fluorocarbonate, polypropylene, PVC, glass or other material with sufficient chemical resistivity and electrical insulating properties. The most important aspect of the construction is perfect separation of the internal space of the electrode from the test solution; this is usually achieved by cementing. Other methods involve sealing of ion-sensitive glass to glass tubing with a suitable thermal dilatation or preparation of heterogeneous membrane electrodes containing polyethylene in the membrane, which is connected to the polyethylene body at high temperature [36]. Mechanical sealing (see e.g. [37]) cannot generally be recommended, as the lifetime and the frequency of faults are not more favourable than those for cemented membranes.

High demands are placed on cements for ISE's. An ideal cement should exhibit high adhesion to the membrane and the electrode body, be hydrophobic (or lyophobic), electro-inactive and electrically non-conductive, have a similar dilatation coefficient to the membrane and electrode body, resist temperature changes and chemical effects of test solutions, not change its volume after application, and remain constantly plastic.

Silicone rubbers vulcanized at laboratory temperature are most often recommended [38, 39]; a two-component cement based on thiokol is also suitable. On the other hand, epoxy resins, waxes, etc. are unsuitable, because they are not sufficiently plastic and may develop cracks. Cementing of LaF_3 crystals is especially difficult; substantial differences in the sensitivity of electrodes prepared in the laboratory were observed when LaF_3 crystals were recemented [40]. An unfavourable effect is apparently exerted by the partially hydrophilic nature of the LaF_3 surface. When the cement around an ISE membrane is cracked, the solution in the cracks is replaced by diffusion only and thus can have a completely different composition from the test solution. This is manifested by slower potential-stabilization, poorer reproducibility and bigger drift. With large cracks the inside and outside of the electrode are short-circuited and the sensitivity and often also the resistance of the ISE decrease.

Solid membranes can be shaped by pressing or be prepared as single-crystals (homogeneous ISE's) or heterogeneous membranes can be made by dispersing the electroactive material in a plastic matrix. The use of more expensive single-crystals (usually prepared by crystallization from the corresponding melt — the Stockbarger method [41]) is still advantage-ous with LaF_3 fluoride-ISE's, as pressed [42] or heterogeneous [43] fluoride-sensitive electrodes generally exhibit poorer properties. Single-crystal membranes can also be made from AgCl and AgBr [18] or from heavy metal chalcogenides [44]. The properties of the latter electrodes depend on small deviations from stoichiometry in the membrane chalcogenide, which are difficult to control during crystallization.

Many inorganic electroactive substances (e.g. AgCl, AgBr or Ag_2S), as well as some organic substances, e.g. radical salts [45], are readily pressed. Pressed membranes can be shaped as required, by using steel containers that can be evacuated (e.g. to 1 mm Hg) and pressures from *ca.* 100 to *ca.* 2×10^4 atm. The pressed pellets can then be sintered; sintering is often carried out during pressing in heated containers [46]. Sintering is indispensable for Pb- and Cd-sensitive ISE's containing Ag_2S + MeS membranes [47].

It is possible to combine various materials and so obtain membranes even from substances pressed with difficulty (e.g. PbS), or increase the conductivity and even change its character in the membrane; a stable electrode response can thus be attained. The best known is the above-mentioned mixture of silver halides or thiocyanate [48] or cupric, lead or cadmium sulphide [49] with silver sulphide, employed by the Orion Research Company for the manufacture of ISE's for halides, SCN^-, Cu^{2+}, Pb^{2+} and Cd^{2+}. Cupric sulphide [50], silver selenide or telluride [51] or silver iodide [52] can also serve as a matrix. A more complicated pressed mixture was used for the preparation of an SO_4^{2-}-ISE, with a membrane containing 31.7 mole % each of Ag_2S, PbS and $PbSO_4$ and 5 mole % of Cu_2S [53]. These electrodes are sometimes classified as heterogeneous, as they contain several components, some of which can be leached by solutions (e.g. AgI from the mixture Ag_2S + AgI in con-centrated solutions of iodide or cyanide). A diffusion barrier is thus formed at the surface, which slows down the response. Occasional polishing with a fine paste is thus necessary; the electrode properties can also be improved by treatment with a silicone oil [1, 54].

Electroactive substances which cannot be pressed or otherwise shaped to form homogeneous membranes can be used in heterogeneous membranes [55, 56] made with various plastics, in which, however, the individual particles must be in contact to maintain electric conductivity. Silicone rubber [56 – 58] (the Pungor electrodes), a thermoplastic polymer [36, 59], dental acrylic [60], epoxide [61], etc. can be used as the matrix. Epoxide and PVC are not suitable for the preparation of heterogeneous Cu_2S membranes [34]. Before use for measurements, heterogenous electrodes mostly require preconditioning by soaking, e.g. in a $10^{-3} M$ solution of the test ion, and their properties may depend on the particle size and the amount of the active substance in the membrane.

For preparation of the best-known Pungor electrodes, a fine-grained precipitate (grain-size from 1 to 50 μm) is mixed with silicone rubber monomer in *ca.* 1 : 1 weight ratio. A hardener and a catalyst are added so that polymerization takes place at laboratory temperature [57].

2.1.5 Liquid Ion-Selective Electrodes (LISE's)

This section includes not only ISE's with liquid electroactive substances or with an electroactive substance dissolved in a suitable solvent (e.g. valinomycin in diphenyl ether), but also those containing the electroactive substance contained in a polymeric film (e.g. PVC) with other components, which are now preferentially used – see Fig. 2.3.

Mechanical problems are inevitable in the manufacture of analytically useful LISE's. The liquid phase containing the electroactive substance must be immiscible with the test solution, loss of the liquid by evaporation should be minimal and the concentration of the electroactive substance in the membrane should be constant. It is evident that these requirements will more probably be met by using liquids with a high viscosity and a low vapour pressure, i.e. usually substances with high molecular weight. LISE's generally place greater demands on the experimenter than do solid membrane ISE's and exhibit poorer selectivity, which, moreover, depends not only on the ion activity ratio, but also on the individual activities of the ions present. On the other hand, they make possible determinations which cannot be carried out with other ISE's, e.g. of Ca^{2+}, NO_3^- and K^+. There are two basic groups of LISE's.

(1) the liquid phase is held mechanically, mostly in a porous membrane, by means of capillary forces;

(2) the electroactive substance is also bound by other forces in a mem-

brane or a film (e.g. electrostatically in a polymeric matrix). The first
analytically useful LISE's were described by Ross [62, 63] and are
commercially available. Esters of phosphoric acid and long-chain aliphatic
amines are used as electroactive substances and are diluted or dissolved
with a suitable solvent; the liquid is kept in a vertical-pore membrane.
Typical porous membranes 7–20 μm thick (millipore filters) are made of
e.g. cellulose acetate or vinyl chloride–vinyl acetate copolymer; a pore
diameter of 10–100 nm is preferable. If the pores are smaller than 10 nm,

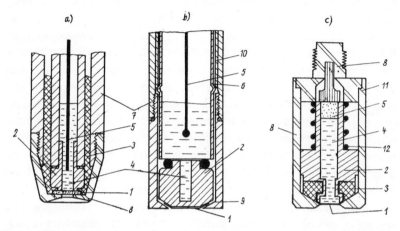

Fig. 2.3 ISE's with membranes containing the electroactive substance in the liquid
phase (LISE). (a) Orion, series 92, (b) LISE with exchangeable sensing element
(according to ref. [3]), (c) Orion, series 93 (the sensing part). 1 — membrane,
2 — clamping piece, 3 — ion-exchanger reservoir, 4 — internal reference solution,
5 — internal reference electrode, 6 — insulating tubing, 7 — the main insulating
body, 8 — bottom cap, 9 — internal flange forming a conical or dished supporting
surface, 10 — electrically conductive tube, 11 — spring.

the liquid is better bound in the membrane, but the selectivity for large
ions may deteriorate (e.g. the Ca^{2+}:Na^+ selectivity is lower by three
orders of magnitude than that obtained with 100-nm pores [62]). In
contrast, too large pores lead to loss of the ion-exchanger. Glass frits
(e.g. 0.5–0.8 mm thick with a maximum pore size of 0.9–1.4 μm) [64]
or porous metal foils (e.g. silver with a pore size of 0.20 μm) [64, 65]
can also serve as membranes. Frits and metals are made hydrophobic
by coating with silicone oils, beeswax, etc. [64]. These membranes
possess the advantage that they are easily connected to the electrode
body. Generally, many materials can be employed for the manufacture
of porous membranes and their selection depends to a certain degree
on the purpose of the application. Recently developed "dropping" ISE's

have the electroactive substance held by capillary forces [65a]. LISE's
with porous membranes have some drawbacks. The pores must be completely filled with the liquid phase, as the existence of air bubbles
(unoccupied pores) can cause short-circuiting between the external
and internal aqueous solution, resulting in potential drift and faulty
performance of the LISE. The solvent is gradually lost by evaporation
and the ion-exchanger slowly dissolves in the test solutions; this limits
the electrode lifetime to some weeks or months. The LISE lifetime is
prolonged by use of an ion-exchanger container in the electrode body
(see Fig. 2.3), which can also be formed by a porous plastic material.
To suppress loss by evaporation, it has been recommended to store
K^+-sensitive ISE's in a closed container with diphenyl ether vapour
[66]. Electroactive substances are more firmly bound and the ISE lifetime is prolonged when they are embedded in a suspension or gel medium. Some membranes used in hydrometallurgical metal separations
[67, 68] can be considered as predecessors of these membrane films,
which can be deposited on a conductive support (see Section 2.1.3) or
placed between two solutions. Inside a membrane film which contains the
polymeric matrix, the electroactive substance and often also a solvent
(the solvent for the polymer is evaporated during preparation of the film)
and/or a plasticizer, coulombic forces can be operative between ions and
any polar sites on the polymer skeleton (on which the electroactive substance can also be adsorbed). The properties of polymeric matrices for
cation-sensitive electrodes are improved by creating a fixed negative
charge on the skeleton [69, 69a]. The method of preparation of the
membrane film, its structure and composition, and the kind of solvent,
plasticizer and polymer can substantially affect the electrode function
and selectivity [70, 71, 71a]. These effects have not yet been completely
elucidated; therefore, recommended procedures must be followed closely.

The plasticizer must reduce the glass-transition temperature of the
polymer and be insoluble in the test solution. Further, the plasticizer
should be a good and non-volatile solvent for the electroactive substance
and maintain (or improve) the selectivity of this substance. Esters of
polybasic acids and nitroaromatics are mostly used as plasticizers. Hence
the esters of phosphoric acid, serving as electroactive substances in
Ca^{2+}-LISE's, can simultaneously function as plasticizers. Similarly to
plasticizers, solvents for electroactive substances are also selected according to the type of substance and compatibility with the other film components. However, building the electroactive substance into a polymeric
film need not necessarily improve the electrode characteristics (see
e.g. [72]).

The most frequently used polymeric matrix in membrane films is PVC [73 – 75, 75a]. The film is prepared by evaporating a solvent (tetrahydrofuran or cyclohexanol) from a mixture of PVC and the electroactive substance (plus a plasticizer and sometimes a solvent for the electroactive substance) on a glass plate. The membrane formed, *ca.* 0.2 mm thick, lasts several months [76].

Many other polymers can be employed, e.g. cellulose triacetate [77] or a hydrophobic elastomeric organopolysiloxane–polycarbonate block copolymer [78]. Another interesting type consists of lipid-containing membranes, e.g. with lecithin [79, 80]. Lipids are natural esters of fatty acids, contained in plant and animal tissues, (i.e. in living membranes), and capable of forming lipophilic gels in which electrostatic forces also operate. Simultaneously, these substances are surface-active, and if a sufficient amount of them is added to the mixture, "solid" membranes can be obtained; these are used e.g. in the electrodes manufactured by the Beckman Company. In contrast to true solid membranes with crystallographic structures, only spatial orientation exists here and their structure resembles that of waxes. For example, a K^+-ISE can be prepared from Nujol, diphenyl ether, lecithin and valinomycin in a ratio of 1 : 1 : 6 : 0.04. (The name "lecithin" is used as a collective term for phosphatides or phospholipids, such as lecithin, lysolecithin, etc.) For preparation of NH_4^+-ISE's, diphenyl ether is replaced by 2-phenyloxydiphenyl or bromodiphenyl ether and valinomycin by nonactin; the ratio employed is 1 : 1 : 6 : 0.1. The NH_4^+-sensitive electrode has improved selectivity, especially with respect to Na^+ [80]. An auxiliary collodion membrane is used to protect commercially available membranes against damage.

Although the overall characteristics of LISE's depend on the solvent [71a], plasticizer and polymeric matrix used, the most important factor is the selection of the electroactive substance itself. There are very many substances at present available; the most important can be classified into four principal groups:

(1) esters of phosphoric acid, used in ISE's for Ca^{2+} and Me^{2+} (Be^{2+}, UO_2^{2+}, Zn^{2+});

(2) neutral ion-carriers (valinomycin, nonactin), employed in K^+-, Ca^{2+}-, NH_4^+-, Ba^{2+}- and Na^+- sensitive ISE's;

(3) organometallic or organoboron compounds of the $(R_a M_n) X_m$ type, where M is a metal or boron atom firmly bound to organic group R (covalent bond) and X is an ion bound at least partly electrostatically and capable of exchange (examples are ISE's for NO_3^-, ClO_4^-, BF_4^- and K^+);

(4) substituted ammonium, phosphonium and arsonium salts, e.g. Aliquat 336 S (methyltricaprylammonium chloride), suitable for the determination of anions, e.g. Cl^-, NO_3^-, and also cations in the form of complex anions, such as $[FeCl_4]^-$ (in which case a constant excess of Cl^- ions must be present in the solution). Other suitable materials involve basic dyes, such as Methylene Blue, Brilliant Green, etc.

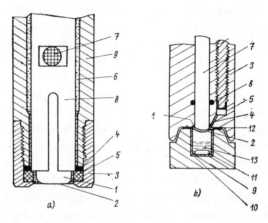

Fig. 2.4 Gas-sensing electrodes. (a) — Orion, series 95, 1 — sensing element, 2 — membrane, 3 — bottom cap, 4 — O-ring, 5 — spacer, 6 — internal filling solution, 7 — reference element, 8 — internal body, 9 — outer body. (b) — air-gap electrode (according to ref. [84]), 1 — surface electrolyte layer, 2 — flat ISE surface, 3 — outer body, 4 — ceramic pin with humidified KCl, 5 — cavity with humidified KCl, 6 — external reference electrode, 7 — ISE, 8 — O-ring, 9 — bottom cap, 10 — sample holder, 11 — magnetic stirrer, 12 — washer.
(b) by permission of Elsevier Publishing Co.

2.1.6 Gas-Sensing and Enzyme Electrodes

By placement of a suitable additional membrane or a film of a suitable solution between an ISE and the sample, certain undesirable effects can be eliminated, e.g. denaturation or haemolysis of blood on contact with a glass ISE [81], or electrodes capable of detecting many non-ionic substances can be obtained (for principles see Chapter 1). These electrodes can be classified as gas-sensing (e.g. for CO_2, NO_2, NH_3, SO_2 and H_2S) and enzyme (e.g. for urea, amygdalin, some amino-acids, etc.) electrodes. The properties of these electrodes strongly depend on the characteristics of the membranes and the fixed solutions and on the electrode construction (Fig. 2.4).

Three kinds of construction are used for gas electrodes.

(a) Heterogeneous membranes, i.e. microporous hydrophobic or lyophobic membranes containing air bubbles inside their pores (air-gap membrane); diffusion across the membrane takes place in the gaseous phase, practically excluding liquid phase transport. A suitable membrane material, which should have up to 60 % "open volume", is polyvinylidene fluoride with an average pore size of less than 1.5 μm [82]; such membranes are employed in NH_3-sensitive electrodes from the Orion Research and EIL companies.

(b) Thin homogeneous films, made of e.g. silicone rubber; the gas transport involves dissolution of the gas in the membrane and consequently the diffusion coefficient is smaller and the response slower (the response-time is generally a very important characteristic of gas-sensing electrodes and determines the practical detection limit [83]).

(c) So-called air-gap electrodes [84] (see Fig. 2.4) contain the electrode surface solution held in a polyurethane sponge above an air bubble separating the test solution and the ISE. The main advantage of this electrode lies in the removal of the danger of membrane wetting (e.g. in the presence of surfactants) or other effects arising from sample contact with the membrane.

The sample preparation for measurement with a gas-sensing electrode involves temperature, ionic strength and osmotic pressure adjustment and, especially, pH adjustment. The pH should have a value such that the solution equilibrium is sufficiently shifted in favour of formation of the gas; i.e. the pH should be two units higher or lower than the appropriate pK value (e.g. pH > 11.2 for the NH_3-sensitive electrode, where $pK_a = 9.25$).

It is also possible to convert only a certain reproducible fraction of the test species into the gaseous state (e.g. with SO_2-sensitive electrodes, where the pK_{a1} value of 1.9 would necessitate addition of too much acid to convert the test species completely into the gas). Care must be exercised to prevent volatilization of the gas. The sensitivity of gas-sensing electrodes depends greatly on the choice of the concentration of the surface solution [85]. From Fig. 2.5 (for the SO_2-sensitive electrode) it is evident that with low $NaHSO_3$ concentrations the sensitivity is higher at low SO_2 concentrations but higher at high SO_2 concentration with high $NaHSO_3$ concentrations [83]. The gas-sensing electrodes described must be differentiated from high-temperature gas electrodes [86], among which the oxygen electrode based on the O_2/O^{2-} couple, reversible on a platinum electrode at temperatures above 500 °C [87, 88], si analytically most significant. This electrode is used for the determination of oxygen dissolved in fused metals and in hot gases; the concentration range is extremely wide (10^{-1}–10^{-20} atm) [89].

In enzyme electrodes [90–92], enzymes function as intermediaries between an ISE and non-ionic compounds, as ions are produced in enzyme-catalysed reactions (see Chapter 1). An enzyme can be added to the measured solution, enzyme reactors can be used [92a], enzyme can be immobilized on the vessel bottom or on the stirrer [93, 94] when

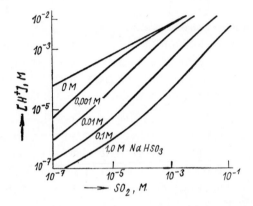

Fig. 2.5 Sulphur dioxide electrode response as a function of internal electrolyte concentration for $pK_{a1} = 1.9$ and $pK_{a2} = 6.8$ (from J. W. Ross, Jr., J. H. Riseman and J. A. Krueger: *Pure Appl. Chem.*, **36**, 473 (1973), by permission of IUPAC).

an air-gap electrode is used, but most electrodes described have the enzyme immobilized as a film on the ISE. The predominantly physical bonding of enzymes in a polyacrylic gel is replaced by chemical bonding (glutaraldehyde + polyacrylic acid) with higher stability. These homogeneous films can be formed directly on the ISE [92] and used without covering them (e.g. with a cellophane membrane), at least in media that are not very septic.

2.1.7 Microelectrodes

In measurements with common ISE's the test solution must have a sufficient volume for immersion of the electrode system (i.e. at least a few ml). Sometimes a sufficient amount of solution is not available (especially in analyses of biological materials or in determinations of ions in geological inclusions, etc.). For smaller sample volumes a modified measuring technique (analyte addition – see Chapter 3) or special electrodes can be used. The simplest way is the use of electrodes with the electroactive material turned upwards; a drop of the sample solution is placed on this surface and a reference electrode is immersed in the drop [95, 96]. Combined electrodes can also be used [97]. These electrodes enable measurements to be made in volumes of some tens of microlitres. For these measurements, so-called mini-electrodes have been constructed [with a diameter of about 1 mm (see e.g. [97a, b])]. For even smaller volumes and especially for intracellular determinations in biology, spe-

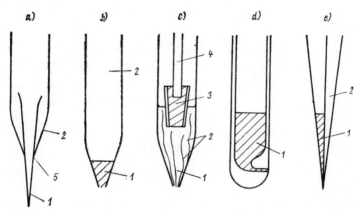

Fig. 2.6 Some types of microelectrode. (a) — glass microelectrode according to Hinke [101]: 1 — cation-sensitive glass microcapillary, 2 — inert glass micropipette. (b) — microelectrode according to Walker, with liquid ion-exchanger [102]: 1 — ion-exchanger, 2 — internal reference solution. (c) — liquid ion-exchanger microelectrode according to Steinhardt [104]: 1 — ion-exchanger, 2 — glass fibres, 3 — porous material with the reference solution, 4 — reference electrode. (d) — side-pore micro-electrode [105, 106]: 1 — ion-exchanger. (e) — double-barrel microelectrode [102, 105]: 1 — ion-exchanger, 2 — reference solution.

cial microelectrodes must be constructed; the construction of glass microelectrodes has been thoroughly investigated (see e.g. [98 – 100]). A typical glass microelectrode is depicted in Fig. 2.6(a) [101]; a glass microelectrode with a solid contact has been described by Riseman and Wall [11]. For various microelectrodes see e.g. [100a].

Microelectrodes with liquid ion-exchangers are mostly made from micropipettes, the tips of which contain the ion-exchanger overlaid by an internal reference solution − see e.g. [102, 103] and Fig. 2.6(b). The main difficulty encountered in preparation of these electrodes is filling of the micropipette tip, with an opening of *ca.* 1 μm, with the ion-exchanger, owing to the hydrophilic character of glass; therefore the micropipette tip is made hydrophobic by a silicone layer. The construction has further been perfected by Steinhardt [104], who inserted several fine glass fibres into the narrowed part of the micropipette, along which the ion-exchanger penetrates into the tip; the electrode body can then be filled with the exchanger from above and an internal reference system can be immersed in it [Fig. 2.6(c)].

Needle-shaped electrodes are especially suitable for intracellular measurements. During measurements outside cells, e.g. in muscles, muscle fibres can be mechanically damaged and the ion-exchanger undesirably stirred during insertion of the electrode. For this purpose, a side-pore microelectrode is suitable [105, 106], Fig. 2.6(d). Reference microelectrodes are usually micropipettes containing the appropriate reference system, placed close to the indicator microelectrode. Sometimes, however, undesirable galvanic cells can be formed in this way in biological materials. Then the use of a double-barrelled electrode is advantageous; one compartment contains the ion-exchanger and the other the reference solution [102, 105], Fig. 2.6(e).

The main problem encountered in measurement with microelectrodes is their high electric resistance ($10^9 - 10^{11}$ Ω), which places great demands on the input impedance of the meter used and on the electrode shielding. In selecting the ISE, electrodes with a poorer performance but a lower resistance are therefore often preferred (e.g. the phenylborate K^+-ISE is often preferred to the valinomycin-ISE). The company Microelectrodes Inc. (Londonderry, USA) manufactures a series of microelectrodes, with diameters from a few µm up to 1 mm for the electroactive part.

An interesting approach to microelectrode construction is a combination of a MOSFET (metal oxide semiconductor field-effect transistor) with an ion-selective membrane which is placed over the MOSFET gate. The electrochemical properties of the MOSFET have recently been studied [107 – 109] and an electrode selective for potassium ions constructed [110]. A MOSFET with a K^+-selective membrane is immersed in the test solution and the drain current is monitored as a function of a_{K^+}. The advantages of this construction are the small dimensions, insensitivity to surfactants and the fact that the measured signal is directly handled electronically in the MOSFET, so that problems with capacity shielding are avoided. The main disadvantages are difficulties with encapsulation of the element in an insulating mantle, and relatively poor stability of the measured signal.

The application of field-effect transistors has been the subject of a recent conference [110a].

2.1.8 Combined Electrodes

Combined electrodes contain both the ISE and the external reference electrode in one body (see Fig. 2.7). This construction has practical advantages and the shielding of the ISE can be used for the lead from the reference electrode. Combined glass electrodes have been used for some time and now combined ISE's in plastic bodies have become common. These electrodes can have a polished surface and be used for analyses of small solution volumes (down to 10 µl) on a planar (e.g. glass) support [97, 111]. The liquid junction is formed by a ground plastic joint or asbestos fibres (Fig. 2.7). Combined ISE's with a built-in thermosensitive resistor [112] or with several membranes for simultaneous determination of the activities of several ions [5] have also been constructed.

Combined ISE's are advantageous in flow-through monitoring, because they exhibit lower streaming potential oscillations than the combination of an ISE and a reference electrode [112a].

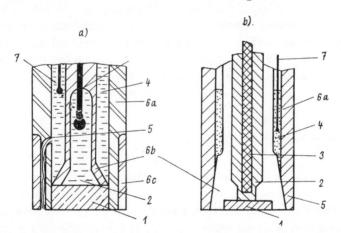

Fig. 2.7 Combined ISE's. (a) — asbestos fibre liquid junction (ref. [111]), (b) — ground plastic joint liquid junction (ref. [97]). 1 — membrane, 2 — internal reference electrolyte, 3 — internal reference electrode, 4 — external reference electrode electrolyte, 5 — liquid junction, 6a — outer electrode body, 6b — internal electrode body, 6c — plastic sleeve fixing asbestos fibres, 7 — external reference electrode.

2.2 REFERENCE ELECTRODES

Potentiometric analysis is based on comparison of the potentials of the indicator and reference electrodes and its precision depends primarily on the stability of these two potentials during the analysis. Problems connected with reference electrodes are sometimes neglected, although they are responsible for a large proportion of faults. The requirements made of reference electrodes can be summarized as follows.

(1) Potential stability.

(2) Potential reproducibility (especially absence of temperature hysteresis).

(3) Low temperature-dependence of the potential and a temperature coefficient of the opposite sign to that of the ISE used.

(4) Universality of application and simple operation.

(5) Low electrical resistance.

(6) Reproducible and small liquid-junction potentials (if such a junction is a part of the reference electrode) and resistance to mechanical blocking; small flow of the reference electrode solution into the test solution.

Any ion-sensitive electrode with a sufficiently stable potential in the test solution can serve as a reference electrode; hence, it can also be an ISE in a solution with a fixed activity of the ions to which it is sensitive. There is no universal reference electrode; the requirements listed are best met by electrodes of the second kind, especially silver chloride, calomel, silver bromide and "thalamide" electrodes. For detailed treatments of the field of reference electrodes, see [113 – 115]; reference electrodes for non-aqueous solvents are reviewed e.g. in ref. [116].

2.2.1 Silver Chloride Electrode

Silver chloride electrodes (see [113 – 118]) are marked by stable and reproducible potentials, low temperature-hysteresis, a wide useful temperature range (up to 275 °C) and easy preparation of electrodes of various shapes. They are often employed as internal reference electrodes in ISE's. They consist of silver chloride deposited on silver or of a mixture of the two substances. The methods of preparation can be classified [117] into: (a) electrolytic, (b) thermal, (c) thermal–electrolytic, (d) other procedures, often involving pressing.

In electrolytic preparation [115, 119], a clean Pt support is silver-plated in a $KAg(CN)_2$ bath and washed, and then 15–25 % of the Ag is converted into AgCl by anodic treatment in $0.1M$ hydrochloric acid. Thermal preparation is based on thermal decomposition of an aqueous paste consisting e.g. of 90% Ag_2O and 10 % $AgClO_3$ [120, 211]; these electrodes are rapidly prepared and have high resistivity, because of the presence of a larger amount of AgCl. As the potential stabilization of thermally prepared electrodes is rather slow, often only the silver layer is prepared thermally, by use of a paste of $Ag_2O + H_2O$, and the silver chloride is prepared electrolytically (the thermal–electrolytic method); in this way work with cyanide is avoided. In analytical work, where the potential-stability of the reference electrode is of prime importance, small differences in the properties of electrodes prepared by various methods (rate of potential-stabilization, small changes in $E°$ after preparation, etc.) are unimportant.

The following preparation procedure for the silver chloride electrode [122] does not satisfy the most rigorous requirements for reproducibility of the electrode properties (the bias potential may attain a value of several mV [114]), but it is simple. A silver wire 0.5–1.0 mm thick is carefully cleaned and degreased with a detergent, rinsed with distilled water and soaked in conc. nitric acid for an hour. After rinsing, the wire is anodically polarized for 30 min by a current of 5–10 mA in $0.1M$ sodium chloride at pH $11-12$ (adjusted with $6M$ NaOH), with a Pt cathode. Too high a current density leads to poorly adhering AgCl films. Several electrodes can be prepared simultaneously by this procedure; electrodes having potentials which differ by more than 0.1 mV from the average value should not be used [122]. When current densities above 12 mA/cm^2 are used, non-porous AgCl layers are obtained [122a].

Freshly prepared silver chloride electrodes exhibit a potential shift of about -0.5 mV during the first $20-30$ hours [123], the time depending on the electrode porosity and the rate of stirring. On the other hand, oxygen dissolved in acidic solutions shifts the potential somewhat towards more positive values; the a_{Cl^-} value in the vicinity of the electrode surface probably decreases because of the reaction

$$2\,Ag + 2\,HCl + 1/2\,O_2 \quad \rightarrow \quad 2\,AgCl + H_2O \qquad (2.1)$$

If a precision of 0.1 mV is required, oxygen should be removed from the solution.

The colour of silver chloride electrodes and their sensitivity to light are still discussed [119, 123, 124]. White, dark grey, pinkish and brownish electrodes have been described; some authors have reported great sensitivity to light, especially in the ultraviolet region. Therefore, large changes in the light should be avoided during measurements, especially direct exposure to sunlight.

Silver chloride electrodes are stored in distilled water or in dilute chloride solutions; they should be handled carefully and the layers must not be touched. Because of increasing solubility of AgCl with increasing KCl concentration and increasing temperature (see Table 2.1), periodic temperature fluctuations may cause gradual dissolution of the film on the electrode. Therefore thermally prepared electrodes and lower chloride concentrations are used for high-temperature measurements.

The temperature hysteresis of silver chloride electrodes is smaller than that of the calomel electrode, but differences of several mV can be observed after temperature variations [125, 126]. Temperature hysteresis can be minimized by constructing the electrode so that the Ag contact touches a solution saturated with both AgCl and KCl at all operational temperatures and the space in which equilibration occurs is small [127].

Table 2.1

Temperature dependence of KCl and AgCl solubilities
and of the solubility of AgCl in KCl solutions at 25 °C

Temperature, °C	0	20	25	50	100
KCl (M)*	3.39	4.02	4.17$^+$	4.80	5.84
KCl (m)	3.70	4.56	4.80	5.71	7.60
AgCl (m)	—	1.05×10^{-5}	—	3.65×10^{-5}	1.46×10^{-4}
KCl conc. (M, 25 °C)		0.0	1.0	4.0	4.17$^+$
AgCl (M)*		1.3×10^{-5}	1.0×10^{-4}	6.4×10^{-3}	7.0×10^{-3}

* From the Foxboro Co. manufacturer's literature
$^+$ A value of 4.16 M is more frequently quoted for saturated KCl at 25 °C [115]

Table 2.2 gives the values of silver chloride electrode standard potentials ($E°$) over a temperature range of 0–90 °C. Differences of the order of tenths of an mV have been observed among $E°$ values determined by various workers; these are ascribed to differences in electrode preparation methods and electrode "history". For any temperature (T; °C), the $E°$ value (V) can be calculated from the equation

$$E° = 0.23735 - 5.3783 \times 10^{-4}T - 2.3728 \times 10^{-6}T^2 \qquad (2.2)$$

with good results up to 200 °C [129].

Analytical chemists are interested in the electrode potentials in chloride solutions of various concentrations rather than in the standard potential (see Table 2.2). As the liquid-junction potential, $\Delta\varphi_L$, is included in these values, they are somewhat dependent on the test-solution composition, and especially on its pH. The $E° + \Delta\varphi_L$ value increases with decreasing

Table 2.2

Potentials of silver halide electrodes (mV) vs. the standard hydrogen electrode

Temperature °C	1 E° AgCl/Ag	2 E° AgCl/Ag	3 E° AgCl/Ag	4 $E^{\circ\prime} + \Delta\varphi_L$ sat. KCl	5 $E^{\circ\prime} + \Delta\varphi_L$ 3.5M KCl	6 E° AgBr/Ag	7 E° AgBr/Ag	8 E° AgI/Ag
0	236.55					81.28		−146.37
5	234.13					79.61		−147.19
10	231.42					77.73		−148.22
15	228.57					75.72		−149.42
20	225.57					73.49		−150.81
25	222.34	222.33		213.8	215.2	71.06		−152.44
30	219.04			208.9	211.7	68.56		−154.05
35	215.65			204.0	208.2	65.85		−155.90
40	212.08			198.9	204.6	63.10		−157.88
45	208.35			193.9	200.9	60.12		−159.98
50	204.49			188.7	197.1	57.04		−162.19
55	200.56			183.5	193.3			
60	196.49	196.8	196.36	165.7*			50.1	
70	187.82	—	187.84					
80	178.7		178.76					
90	169.5	169.6	169.26			—	25.1	—
Ref.	[128]	[129]	[130]	[115]	[115]	[132]	[133]	[134]

* Ref. [137].

$E^{\circ\prime}$ involves the E° of the cell plus a term containing γ_{Cl^-} (KCl) owing to the dependence of $\Delta\varphi_L$ on the test-solution composition, the $E^{\circ\prime} + \Delta\varphi_L$ value also varies somewhat. Values of $E^0_{(25\,°C)} = +221.59 \pm 0.72$ mV, $E^0_{(50\,°C)} = +204.00 \pm 0.82$ mV and $E^0_{(75\,°C)} = +184.71 \pm 0.91$ mV were obtained with a membrane AgCl-ISE [134a].

pH (at pH < 2 it is at least *ca*. 1 mV more positive and at pH > 12 at least *ca*. 1 mV more negative than in neutral solutions). In addition to the saturated silver chloride electrode, electrodes with 4.0, 3.5, 1 and 0.1M KCl are also used (NaCl is not suitable at concentrations above 2M, because solid solutions with AgCl are formed [135]). For electrodes with 1 and 0.1M KCl, potential values of 236 and 289.5 mV, respectively, are reported [136].

2.2.2 Calomel Electrode

Calomel electrodes [125, 138 − 140] are still the most popular external reference electrodes. They consist of mercury covered with a paste of Hg and Hg_2Cl_2, in contact with a chloride solution (usually KCl, less frequently NaCl or LiCl). For the saturated calomel electrode (SCE), the damp paste can be prepared with KCl crystals; solid KCl is also present in the solution (saturated KCl solution, 4.16M or 4.80m at 25 °C [115]). With SCE's it is, of course, not necessary to control the KCl concentration carefully, but there is a danger of separation of KCl crystals in some types of liquid junction, owing to variations in KCl solubility with temperature (see Table 2.1). Therefore, a calomel electrode with 3.5M KCl is sometimes recommended; electrodes with 0.1 and 1M KCl were formerly used frequently, but now are rarely met in the laboratory. A disadvantage of classical calomel electrodes is their considerable temperature hysteresis. If the electrode is exposed to a temperature change of 10 °C or more, its potential is not completely stabilized for several hours. The change in potential from the equilibrium value increases with increasing temperature change and the time required for restoration of the equilibrium decreases with increasing temperature. SCE hysteresis is characterized by a small initial potential shift away from the equilibrium value, followed by an exponential approach towards it. The initial shift is larger and the stabilization time longer when the change is from higher to lower temperatures (see Fig. 2.8. and refs. [141, 142]). The chief cause of true hysteresis (potential changes due to departure of the whole cell from temperature equilibrium are also sometimes taken as hysteresis) is insufficient reversibility of the disproportionation reaction

$$Hg_2Cl_2 \rightleftharpoons Hg + HgCl_2 \qquad (2.3)$$

As the rate of reaction (2.3) increases with increasing temperature, the mercuric salt remains in the system even after a decrease in the temperature [143]. Calomel electrodes are not recommended for prolonged use

Table 2.3

Potentials of calomel electrodes (mV) vs. the standard hydrogen electrode

Temperature °C	$E°$				$E°' + \Delta\varphi_L$			
	1	2	3	4	5	6	7	8
	Hg_2Cl_2/Hg	Hg_2Cl_2/Hg	Hg_2Cl_2/Hg	Hg_2Cl_2/Hg	sat. KCl, Hg_2Cl_2/Hg	sat. KCl, Hg_2Cl_2/Hg	4.0M KCl	3.5M KCl
0	274.0				259.2			
5	273.1	272.9	272.90		—			
10	—	271.9	271.92		253.9	254.3	—	255.6
15	270.9	270.8	270.86		—	251.1	—	—
20	—	269.6	269.55		247.7	247.9	—	252.0
25	268.0	268.2	268.18	268.05	244.5	244.4	245.9	250.1
30	—	266.6	266.59		241.2	241.1	243.8	248.1
35	265.0	264.9	264.85		—	237.6	—	—
38	—	—	—		234.5	235.5	240.2	244.8
40	—	263.0	263.02		—	234.0	239.3	243.9
45	260.5	261.0	261.01		227.4			
50	—							
55	256.1							
70				249.55				
100				232.95				
Ref.	[114]	[144]	[140]	[125]	[137]	[115]	[115]	[115]

The values in column 5 are rounded; the values in column 4 were calculated by using the data from ref. [129].

at temperatures higher than 70 °C (although they have been used at up to 200 °C [125]), since disproportionation and the dissolution of calomel by formation of chloro-complexes then proceeds at such a rate that the potential can become unstable and the electrodes have a shorter life [141]. By suitable modification (especially by making the space in which the equilibrium is established small), hysteresis effects can be substantially suppressed [125, 139, 140, 143].

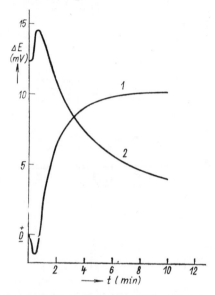

Fig. 2.8 Saturated calomel electrode hysteresis. (Orientative determination, K 4018 type electrode, Radiometer; two identical electrodes were connected without liquid junction). Temperature change: (1) — 20 → 60 °C, (2) — 60 → 20 °C. Curve (2) was obtained 180 min after curve (1).

Calomel electrodes are also somewhat sensitive to the presence of oxygen in acidic solutions, but the reaction

$$2\,Hg + 2\,HCl + 1/2\,O_2 \;\rightleftharpoons\; Hg_2Cl_2 + H_2O \tag{2.4}$$

explains this effect only partially, as the effect of low concentrations of oxygen is reversible [114].

2.2.3 Thallium Amalgam—Thallous Chloride ("Thalamide") Electrode

This electrode (manufactured by Jenaer Glaswerk Schott, Mains, GFR) consists of a 40 % thallium amalgam in contact with a saturated solution of thallous chloride and a KCl or NH$_4$Cl solution. It can be employed at temperatures from 0 to 135 °C and its main advantage is negligible temperature hysteresis [126]. The electrode potential depends on the Tl content in the amalgam: for 40 % w/w at 25 °C, $E°$ (without $\Delta\varphi_L$) is -578.4 mV [145], Table 2.4. This electrode has so far been rarely used, although it is considered promising.

Table 2.4

Potentials of thalamide electrodes (mV) vs. the standard hydrogen electrode

Temperature °C	$E°' + \Delta\varphi_L$ sat. KCl, TlCl/Tl/Hg)	$E°' + \Delta\varphi_L$ 3.5 m KCl
5	-562.4	-561.0
10	-565.2	-563.5
15	-568.7	-566.0
20	-572.7	-568.6
25	-576.7	-571.3
30	-580.6	-574.3
35	-584.6	-576.7
40	-588.9	-579.6
50	-597.1	-585.4
60	-605.7	-591.6
70	-614.4	-598.0
80	-622.9	-604.8
90	-630.9	-611.9
Ref.	[145]	[146]

2.2.4 Other Reference Electrodes

Silver bromide and iodide electrodes [123] can be prepared similarly to silver chloride electrodes; however, they are not often used. Though the silver bromide electrode has a reproducibility comparable with that of silver chloride electrodes and is suitable even for high temperatures [133], the reproducibility of the silver iodide electrode potential is poorer (for the $E°$ values see Table 2.2).

The mercurous sulphate electrode [114, 147, 148], $Hg/Hg_2SO_4/sat.$ K_2SO_4, is similar to calomel electrodes. Its potential is unfavourably affected by larger and more variable liquid-junction potentials during measurements, but it has lower temperature hysteresis (probably because the solubility of Hg_2SO_4 is greater than that of Hg_2Cl_2). However, it is not recommended for temperatures above 70 °C. The $E° + \Delta\varphi_L$ potential of the electrode containing saturated K_2SO_4 solution is *ca.* $+658$ mV (according to ref. [149]). This electrode is mostly applied for measurements where chloride ions flowing from a silver chloride or calomel electrode would interfere (e.g. in the determination of chloride). Such measurements can also be carried out with any reference electrode with two liquid junctions. However, the bridge electrolyte must be periodically replaced.

2.3 LIQUID JUNCTION

2.3.1 Introduction

Four basic cell types can be used for ISE measurements:

(*a*) $\underbrace{\text{Ag} \mid \text{AgBr} \mid x\text{Br}^-, y\text{F}^- \mid \text{LaF}_3}_{\text{F}^-\text{-ISE}} \mid \text{test solution} \mid \text{AgCl} \mid \text{Ag}$

$\text{Cl}^- = \text{const.}$ (2.5)

(cell without liquid junction)

(*b*) $\underbrace{\text{Ag} \mid \text{AgBr} \mid x\text{Br}^-, y\text{F}^- \mid \text{LaF}_3}_{\text{F}^-\text{-ISE}} \mid \text{test solution} \parallel \text{KCl sat.} \mid \text{Hg}_2\text{Cl}_2 \mid \text{Hg}$ (2.6)

(cell with one liquid junction)

(*c*) $\underbrace{\text{Ag} \mid \text{AgCl}}_{\text{Cl}^-\text{-ISE}} \mid \text{test solution} \parallel 3M\ \text{NH}_4\text{NO}_3 \parallel 1M\ \text{KCl} \mid \text{AgCl} \mid \text{Ag}$ (2.7)

(cell with two liquid junctions)

(*d*) $\underbrace{\text{Ag} \mid \text{AgCl} \mid x\text{Cl}^-, y\text{F}^- \mid \text{LaF}_3}_{\text{F}^-\text{-ISE}} \mid \text{test} \parallel s\text{F}^- \mid \underbrace{\text{LaF}_3 \mid y\text{F}^-, x\text{Cl}^- \mid \text{AgCl} \mid \text{Ag}}_{\text{F}^-\text{-ISE}}$

soln. (2.8)

(cell with two ISE's and one liquid junction)

Cells (*b*), (*c*) and (*d*) contain either one liquid junction or two, which can create difficulties because of the formation of liquid-junction potentials (see Chapter 1) and because of possibility of mechanical blocking by KCl crystallization, adsorption of organic substances from biological solutions, formation of $KClO_4$, etc. Cell (*a*) contains no liquid junction but its use in analysis is limited, as it presupposes identical activities of the ion determining the external reference electrode potential, in all test solutions and standards. Errors caused by variations in this activity may exceed those caused by the liquid-junction potential.

The liquid-junction potential, $\Delta\varphi_L$, depends primarily on differences in the pH and the solvents [150] of the adjoining solutions, and also on differences in concentration of all ions, on the junction type and on temperature (i.e. on all factors affecting the mobilities of charged species). Temperature differences of several degrees between the standard and the sample cause no substantial changes in $\Delta\varphi_L$ [125, 149], but at higher temperature differences (e.g. when two H_2SO_4 solutions are contacted), thermal diffusion effects arise.

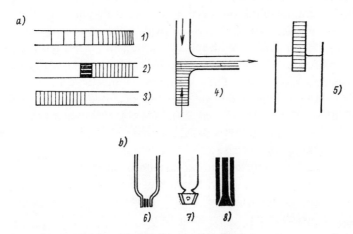

Fig. 2.9 Liquid junction types. (A) Guggenheim types (ref. [151]): 1 — continuous mixture, 2 — constrained diffusion, 3 — free diffusion, 4 — flowing junction, 5 — "indefinite" junction. (B) Some common liquid junction types: 6 — ceramic liquid junction, 7 — glass sleeve junction, 8 — plastic sleeve junction.

2.3.2 Types of Liquid Junction

Guggenheim [151] defined and experimentally studied several types of liquid junction: (a) continuous-mixture junction; (b) free-diffusion junction; (c) constrained-diffusion junction; (d) sharp junction; (e) flowing junction (see Fig. 2.9). The individual junction types exhibit somewhat different $\Delta\varphi_L$ values (the differences amount to tenths of an mV, provided that cylindrical symmetry of the junction is preserved, but may be several mV if a thinner tube is immersed in a large vessel [151, 152]. The junction types also differ in their resistance to blocking, in the rate of attainment of steady-state diffusion and in the electrical resistance. Most liquid junctions of commercial reference electrodes are a compromise between the requirements for an ideal junction and for mechanical strength and practical usefulness; some of them (for a survey see [153]) are difficult to classify among the Guggenheim types.

A free-diffusion junction consists of direct contact of two solutions, e.g. in a capillary with a diameter of less than 1 mm, the air being removed by pipette [154]. The $\Delta\varphi_L$ value is considered to be most stable here (changes of hundredths of an mV per hour), and blocking is infrequent, but mutual contamination of the solutions can be considerable. Another type of junction can be obtained by shifting glass or "Plexiglas" plates against one another [154].

The constrained-diffusion junction (a porous diaphragm) is probably the most practically useful type, provided that frequent mechanical blocking is not a danger. With this type of junction good reproducibility and stability can also be obtained [154], although biased fluctuations are usually encountered. The diaphragm is mostly ceramic or glass (see e.g. [155 – 157]); the flow from the electrode is very small, less than 1 µl/hour for a 10-cm liquid column [141]. Unstable $\Delta\varphi_L$ values can be observed at extreme pH values; in strongly alkaline solutions these junctions are poorly resistive.

Liquid junctions with asbestos fibres embedded in soft glass [158, 159] are no longer used frequently, because of rapid drying of the junction and the marked dependence of its properties on the preparation procedure, which is poorly reproducible.

A porous junction is also formed by sealing Pd wire(s) in glass. The flow of solution can be extremely small, but the redox sensitivity is a disadvantage; this is also observed with porous graphite diaphragms [160]. A Pd wire can be sealed in glass with a slightly different coefficient of dilatation, so that a crevice is formed on cooling. Porous diaphragms made of two glasses of different dilatation coefficients are based on a similar principle.

Reference electrodes with junctions formed by hydrophilic or hydrophilized plastics [161, 162] (e.g. PTFE with 15 % w/w of glass fibres) are used for biological measurements in blood, serum, plasma and other viscous liquids and suspensions. Sometimes surfactants are added to the solution to enhance the junction permeability for water. Transport across these membranes is mostly given by diffusion and therefore the potential stability is unaffected by flow from the electrode, as with other porous junctions. Of course, the diaphragm material must not contain ionogenic groups; otherwise an additional membrane potential would develop. For this reason, diaphragms made from ion-exchange resins are quite unsuitable.

Ground-glass junctions are recommended for emulsions, solutions of proteins, detergents and for media with high electrical resistance [163]. The junction is formed by a film of electrolyte in a roughly finished ground-glass joint, or a channel is made in the joint and the solutions are connected by turning the two parts of the joint to a certain position. An advantage is that the junction can readily be taken apart and a blocked junction quickly freed. On the other hand, crystals of KCl can push the two parts of the joint apart and increase the flow from the electrode, already relatively high, even more.

A flowing junction is formed by continuous laminar flow and uniform mixing of two solutions. Although stable $\Delta \varphi_L$ values can be obtained, this arrangement is not suitable for practical measurements, except for continuous flow-through analysis.

Reference electrodes with conductive membranes made from non-porous carbon, silicon or germanium [164] or manganese and iron oxides [165] do not attain a stability equal to that of common reference electrodes and their potential is not quite independent of the solution properties.

2.3.3 Minimization of Errors due to the Liquid-Junction Potential

In an analytical laboratory it is first necessary to prevent gross errors due to blocking of the liquid junction, by checking it periodically. KCl crystals are removed by immersion in hot distilled water: if a precipitate of known composition is present, it is dissolved by using suitable complexing agents or, if possible, the junction is taken apart and cleaned. Errors arising from the unavoidable existence of $\Delta \varphi_L$ are minimized (a) by suitable selection or adjustment of the solutions or by including a bridge between the test solution and the reference electrode internal solution and (b) by correcting for $\Delta \varphi_L$ by calculation or extrapolation.

The residual liquid-junction potential in measurement of samples and standards can be suppressed by a suitable cell arrangement, so that it is negligible with respect to the overall potential change. If an identical indifferent electrolyte is added in sufficient concentration on both sides of the liquid junction, the activity coefficients of the variable components will be determined virtually by the character and concentration of this electrolyte [166] and will consequently have similar values on both sides of the junction. The transport numbers will also be similar. Most frequently a solution of constant composition will be on one side of the junction (the internal solution of the reference electrode) and the composition of the test and standard solutions is adjusted to a similar ionic strength and pH. Small differences in the residual $\Delta\varphi_L$ still exist, but with a suitable type of liquid junction and a concentrated solution in the reference electrode, the $\Delta\varphi_L$ value is reproducible in practice. If measurements are carried out on solutions with pH values outside the range *ca.* 2.5–11.5, the reference electrode internal solution should be brought to a similar pH.

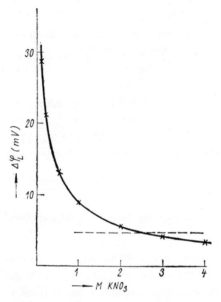

Fig. 2.10 Dependence of $\Delta\varphi_L$ on the potassium nitrate concentration in the liquid bridge for the 0.1N HCl//xM KNO$_3$//sat. KCl system. The values were calculated from the Henderson equation; the value for the junction 0.1N HCl//sat. KCl (+4.6 mV) is given as the dashed line.

In order to reduce $\Delta\varphi_L$, a cell with two liquid junctions can be used, provided that the electrolyte concentration in the bridge is substantially larger than the concentrations in the adjoining solutions; then $\Delta\varphi_L$ is virtually determined by the composition of the bridge electrolyte and the two $\Delta\varphi_L$ values will be similar but with opposite signs and so will compensate one another [115, 151] (see Fig. 2.10). In Fig. 2.10, the calculated dependence of $\Delta\varphi_L$ on the concentration of KNO$_3$ in the bridge is depicted for the system

$$0.1 M \text{ HCl} \parallel xM \text{ KNO}_3 \parallel \text{sat. KCl}$$

It is evident that when one of the adjoining solutions is concentrated, the introduction of the bridge does not lead to a decrease in $\Delta\varphi_L$ and actually for a KNO_3 concentration less than *ca.* $2.5M$, the overall $\Delta\varphi_L$ is greater than the $\Delta\varphi_L$ value for the junction $0.1M$ HCl || sat. KCl (i.e. *ca.* $+4.6$ mV).

The bridge solution should contain cations and anions of similar mobilities; $3.5M$ or saturated solutions of KCl, or solutions of mixtures of KCl and KNO_3, and concentrated solutions of NH_4NO_3 and KNO_3 are most often used. For practical reasons, a liquid bridge is usually included in reference electrodes (double-junction reference electrodes); earlier, external bridges, mostly agar, were often used, but they dry out very quickly.

2.3.4 Calculation of Liquid-Junction Potentials

An equation for the liquid-junction potential was given in Chapter 1, based on the simplifying assumption that activities are equal to concentrations. Generally it holds that

$$\Delta\varphi_L = -\frac{RT}{F} \int_1^2 \sum \frac{t_i}{z_i} \, d \ln c_i - \frac{RT}{F} \int_1^2 \sum \frac{t_i}{z_i} \, d \ln \gamma_i \qquad (2.9)$$

where t is the transport number. That is:

$$\Delta\varphi_L = (\Delta\varphi_L)_c + (\Delta\varphi_L)_\gamma \qquad (2.10)$$

While the first term on the right-hand side can be calculated given certain assumptions concerning the interface structure (see Chapter 1), the assessment of the second term is difficult [114, 115, 152, 167]. However, its value is mostly substantially smaller than $(\Delta\varphi_L)_c$, except for certain cases, e.g. the interface $0.01M$ HCl || $0.1M$ KCl, where $(\Delta\varphi_L)_\gamma$ may amount to as much as 30 % of $\Delta\varphi_L$ [152]. $\Delta\varphi_L$ is usually assessed from the Henderson equation [see Eq. (1.7)], derived for the continuous-mixture type junction, and considering only $(\Delta\varphi_L)_c$.

Some values of $\Delta\varphi_L$ are given in Table 2.5. They were calculated from the Henderson equation in the form

$$(\Delta\varphi_L)_c = \frac{RT}{F} \frac{(u_1 - v_1) - (u_2 - v_2)}{(u_1' + v_1') - (u_2' + v_2')} \ln \frac{u_1' + v_1'}{u_2' + v_2'} \qquad (2.11)$$

with u, v, u' and v' defined by

$$u = \Sigma c_+ \lambda_+^0; \quad v = \Sigma c_- \lambda_-^0; \quad u' = \Sigma c_+ \lambda_+^0 \, |z|; \quad v' = \Sigma c_- \lambda_-^0 \, |z|$$

where the λ^0's are limiting equivalent conductivities. For the most frequent case, when one of the solutions is saturated KCl, and for 25 °C,

Eq. (2.11) gives for the junction potential (in volts):

$$\Delta(\varphi_L)_c = 0.0592 \frac{u_1 - v_1 + 11.9}{u_1' + v_1' - 623} \log \frac{u_1' + v_1'}{623} \qquad (2.12)$$

(sat. KCl $= 4.16M$; $\lambda_{K^+}^0 = 73.50$, $\lambda_{Cl^-}^0 = 76.35$).

In the literature, good agreement of the values calculated from the Henderson equation with experimental values is reported [169, 170] (this is perhaps due to mutual compensation of errors arising from various simplifications), but in some works it is stated that this assessment is of limited value [1]. The agreement is better for interfaces involving ions of the same valence, e.g. KI/KCl (an error of *ca*. 1–2 mV [113]), than for interfaces such as e.g. KCl/K$_2$SO$_4$, where a difference of 5.3 mV was found when the concentrations were 0.1M/0.05M [171]. In analytical measurements these calculations are rarely used, because of their tediousness and relatively poor precision. However, the $\Delta\varphi_L$ value as well as the residual $\Delta\varphi_L$ should be considered when selecting the experimental arrangement.

Table 2.5

Liquid-junction potentials (mV) calculated from Eq. (2.11) for 25 °C

1M KNO$_3$: sat. KCl	+1.1	1M HNO$_3$: sat. KCl	+14.4
0.1M KNO$_3$: sat. KCl	+1.9	0.1M H$_2$SO$_4$: sat. KCl	+7.0
10^{-2}M KNO$_3$: sat. KCl	+2.9	10^{-2}M NaOH : sat. KCl	+2.3
10^{-3}M KNO$_3$: sat. KCl	+4.1	0.1M NaOH : sat. KCl	—0.4
10^{-7}M KNO$_3$: sat. KCl	+7.5	1M NaOH : sat. KCl	—8.6
0.1M NaNO$_3$: sat. KCl	+1.6	1M KOH : sat. KCl	—6.9
0.09M NaNO$_3$: sat. KCl	+1.6	1M Ca(NO$_3$)$_2$: sat. KCl	—4.4
+0.01M NaF		0.1M Ca(NO$_3$)$_2$: sat. KCl	+4.7
0.1M NaF : sat. KCl	+1.9	1M HCl : 3.5M KCl	+15.4
1M HCl : sat. KCl	+14.1	1M HCl : 1M KCl	+26.9
0.1M HCl : sat. KCl	+4.6	1M HCl : 0.1M KCl	+57.3
10^{-2}M HCl : sat. KCl	+3.0	0.1M HCl : 0.1M KCl	+26.9

The + sign denotes that the polarity is X$^-$: $^+$KCl; the λ^0 values used in the calculation were taken from ref. [131].

Note: For the EMF of the measured cell, e.g.

$$^-Hg \mid Hg_2Cl_2 \mid \text{sat. KCl}^+ \mid\mid {}^-0.1M \text{ NaF} \mid LaF_3 \text{ - ISE}^+$$
$$E_{ref.} \qquad \Delta\varphi_L \qquad E_{ISE}$$
$$E = E_{ISE} + \Delta\varphi_L - E_{ref.}$$

Hence, if the sign of $\Delta\varphi_L$ is +, $\Delta\varphi_L$ must be added to the E_c value for correction; negative $\Delta\varphi_L$ values are subtracted.

2.3.5 Suspension Effect

This effect sometimes causes considerable differences in the potentials measured in a suspension or colloid and in the supernatant liquid or filtrate. The sign of the difference, $\Delta pH(\Delta pX) = pH_{susp.}(pX_{susp.})$ $- pH_{soln.}(pX_{soln.})$, depends on the charge on the solid particles; for negatively charged particles, $pH_{susp.} < pH_{soln.}$ [172]. With uncharged suspensions, $\Delta pH(\Delta pX) = 0$. There is a lack of data concerning the suspension effect with ISE's other than the glass electrode.

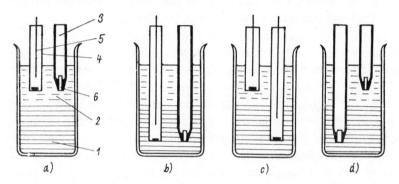

Fig. 2.11 The suspension effect. 1 — suspension, 2 — supernatant liquid, 3 — external reference electrode, 4 — ISE, 5 — internal reference electrode, 6 — liquid junction. In the presence of the suspension effect, the results obtained in positions *a* and *b* are different. A potential difference is obtained in position *d*, but not in position *c*.

The cause of this effect has not been explained unambiguously; it is assumed that a very high liquid-junction potential is formed, either because of the effect of colloidal particles on the relative mobilities of K^+ and Cl^-, or of formation of a liquid-junction potential at the suspension/supernatant-liquid interface. Another possibility is the formation of a Donnan potential at the latter interface, where colloidal particles with ion-exchanging properties function as a semipermeable membrane.

The position of the reference electrode, but not that of the ISE, is important for the suspension effect, as follows from Fig. 2.11 [173]. This effect must be considered in analyses of soil extracts, blood, plasma, etc., which contain charged colloids; in extreme cases the ΔpX value can amount to several units. According to LaMer [174] the effective value of pH (pX) corresponds to that for the localized part of the solution between the charged particles of the slurry.

2.4 MEASURING INSTRUMENTS

2.4.1 Activity Meters

The selection of a meter is chiefly guided by the precision of measurement required, the application intended and the maximum impedance of the cell. Many manufacturers produce meters useful for work with ISE's; among the companies with the widest range are Beckman Instruments, Corning-EEL Scientific Instruments, Metrohm, Orion Research, Philips and Radiometer. The Foxboro and Uniloc companies produce industrial and continuous measurement instruments.

Activity meters can be classified into two groups: (*a*) direct-reading instruments (current less than 10^{-11} A passes through the cell); (*b*) compensation meters, where the measured voltage is compensated by a standard voltage, so that virtually no current passes through the cell.

High precision can be attained with instruments in the second group, but work with them is tedious and they cannot be connected to recorders. For this reason, direct-reading instruments, which can be connected to a suitable recording instrument and to a printer, are usually preferred. Meters with an analogue or digital logarithmic display (e.g. the Orion Research 407A Specific Ion Meter or Philips PW 9414) provide, among other things, direct reading in mg, ppm, %, etc., without use of calibration curves, provided that in the given activity range E is a linear function of a with slope S. Unfortunately, it is the E vs. a range close to the detection limit, where the dependence is not linear, that is often of analytical interest; here gross errors can be committed by using these instruments, increasing with increasing dependence of S on a and with increasing difference in the activities of the standard and the sample. A new generation of instruments is represented by the microprocessor-controlled ion-meter Orion Model 901, which evaluates the results automatically by calibration curve or addition techniques, corrects for the background value and makes the determination substantially easier.

If high precision is not required, instruments with scales divided to 1 mV can be employed; however, a precision of 0.1 mV is usually necessary. Instruments differentiating pH only to 0.1 (*ca.* 6 mV) are quite unsuitable, as an error of 1 mV with an ISE exhibiting Nernstian response corresponds to a relative error of 3.9 % for univalent ions and 7.8 % for bivalent ions (at 25 °C; see Fig. 2.12). The meters are based on precise d.c. voltmeters with high stability, adapted for measurements on cells with especially high resistances. The input resistance of the meter must be at least 10^3 times that of the cell resistance, in order that the error

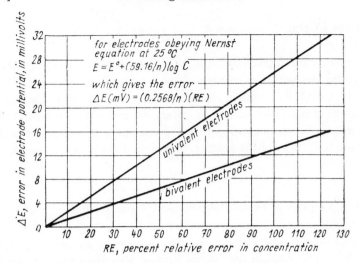

Fig. 2.12 Theoretical error in potential as a function of relative error in concentration. (Foxboro-Yoxall manufacturer's literature).

should not exceed 0.1 %. With modern instruments (an input impedance of $> 10^{13}\ \Omega$) a reproducibility of 0.05 mV (i.e. *ca.* 0.001 for pX) can be attained. Hence potential-fluctuations caused by thermal non-equilibrium in the cell or by variations in the ISE characteristics are usually much higher than the instrumental fluctuations.

2.4.2 Automatic Titration

Instruments for automated titrations have also been developing rapidly. For example, the Radiometer RTS 622 is suitable for automation of all kinds of potentiometric titrations [175]. A precision of 0.1 % is attained for sharp end-points. Even poorly developed titration curves can yield precise results when the data are analysed by computer [176]. The apparatus consists of a digital pH/mV-meter, an automatic titrator, a precise autoburette and a recorder adapted for the purpose. The apparatus is of the module type, so that its individual components can be used independently.

For series of titrations cheaper instruments can be used, e.g. the Radiometer ETS 611, with which no recorder is employed and the titration is carried out to a preset end-point. When the Radiometer DTS 634 apparatus (Digital Titration System) is used, the end-point is determined by a built-in microcomputer from the second derivative of the titration curve. With this instrument 60 titrations can be done per hour. Other titration instruments are manufactured by the Metrohm and Beckman companies, etc.

2.4.3 The pX-stat

The activity of a given ion in solution can be kept constant by adding controlled amounts of a solution of the ion, by using a pX-stat. This is actually a kind of automatic titration, in which the potential corresponding to the required ion-activity is preset, instead of the end-point. The activity can be maintained within a given range by using two simple pX-stats and titrating the solution with a solution of the given ion and with distilled water.

2.4.4. Automation of Direct Potentiometric Determinations

Direct potentiometric measurements can also be automated (see e.g. [177, 178]) and instruments for this purpose are commercially available, e.g. the Stat/Ion from the Photovolt Corporation company, which simultaneously analyses blood or urine for Na, K and Cl by direct potentiometry and CO_2 by coulometric titration with electrogenerated OH^-. Up to 48 samples per hour can be analysed. The Radiometer ABL 1 Acid Base Laboratory instrument is designed for the determination of pH, pCO_2 and pO_2 in blood. Automatic correction is made for the barometric pressure and the haemoglobin concentration and the results can be printed directly on cards suitable for filing in hospital records. A similar instrument is the automated analyser for ionized calcium in blood, the Orion Biomedical Model SS-20, which can make a correction for the sodium content. With any of these instruments, an analysis does not take more than 5 min and the required sample volume is 0.5 ml. The Orion Research Company has also developed the Space-Stat instrument, determining pH, pCO_2, Na^+, K^+ and the overall and ionized Ca in blood and urine, for astronautical purposes [179].

2.4.5 Electrostatic Noise and Shielding

Measurements are frequently subject to considerable noise, especially when variable electromagnetic fields are present (e.g. close to electrical machines, etc.). During measurements on cells with very high impedance (e.g. gas-sensing electrodes or microelectrodes with an impedance of 10^9–10^{10} Ω; for resistance measurement see [180, 181]), considerable noise can be generated even by a.c. lines in the building. The noise can be decreased somewhat by using special soldering materials for the contacts. With high-resistance electrodes, an IR drop may also appear (e.g. 2 mV for a current of 1×10^{-12}A and 2000 MΩ), but for most electrodes with a resistance below 100 MΩ this is negligible.

If the ISE has a substantially higher resistance than the reference electrode, the noise is carried along the lead of the reference electrode and the measurement is virtually unaffected. If, however, the resistance between the ISE and the reference electrode is similar to the input impedance of the meter, a voltage divider is formed (see Fig. 2.13) and the electrode sensitivity decreases. If the insulation resistance between the test solution and the earth is not at least 100 times that between the reference electrode and the earth, then the reference electrode potential and the buffer adjustment setting on the meter are seriously affected (see Fig. 2.13).

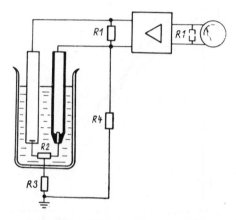

Fig. 2.13 Division of electrode chain potential and earth error. If resistances R_1, R_3 or R_4 are not substantially larger than R_2, the measurement is subject to an error.

All parts of the instrument should be earthed to one point. Long leads and coils on leads should be avoided, especially when large resistances are involved. The insulation is also affected by the air humidity (20–60 % humidity at laboratory temperature is ideal). At low humidities the danger of electrostatic faults increases, while at high humidities the electrical leakage increases.

2.5 EFFECT OF TEMPERATURE

The simplest equation for the measured ISE potential is

$$E = E_c^0 + \frac{RT}{zF} 2.303 \log a_i \qquad (2.13)$$

The dependence of the potential (in mV) on the temperature is then obtained by differentiating Eq. (2.13) and substituting for $2.303R/F$ [182, 183]:

$$\frac{\mathrm{d}E}{\mathrm{d}T} = \frac{\mathrm{d}E_c^0}{\mathrm{d}T} + \frac{0.1984}{z}\log a_i + \frac{0.1984T}{z}\frac{\mathrm{d}\log a_i}{\mathrm{d}T} \qquad (2.14)$$

Changes in $\Delta\varphi_L$ with temperature can usually be neglected. The third term on the right-hand side of Eq. (2.14) (the solution temperature-coefficient) expresses the changes in the activity with the temperature, due to changes in the activity coefficients and in equilibrium constants and small changes

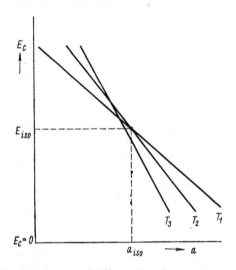

Fig. 2.14 Isopotential (for explanation see the text).

in the activities because of changes in the solution volume with temperature. The second term (the slope temperature-coefficient) expresses changes in the ISE slope with temperature and is often incorrectly considered to be the only cause of changes in the cell EMF with temperature. The Nernstian slope values for some temperatures are given in the appendices (Table A); generally the slope value can be calculated from the relation $S = 0.1984T/z$ (mV/deg), i.e. S changes by $ca.$ 1 mV for a temperature change of 5 °C if $z = 1$. The first term (the standard cell temperature-coefficient) is characteristic of the given cell and depends on its composition and the electrode construction; it is generally advantageous when it is as small as possible. If interferences from other ions are considered, the temperature coefficient, $\mathrm{d}\ln k_{ij}^{Pot}/\mathrm{d}T$, must be borne in mind, i.e. the electrode selectivity changes with a change in the temperature. This temperature coefficient arises from the effect of temperature change on the enthalpies of the ion transport and of the appropriate exchange reaction between the ions. So far, no data are available from which this coefficient could be theoretically calculated.

Sometimes it is difficult to standardize and measure at the same temperature (which gives the best precision). In that case it is possible, however, to employ the isopotential technique [184, 185]. If E vs. a is plotted for two different temperatures, e.g. T_1 and $T_2 = T_1 + 20\ ^\circ C$ (see Fig. 2.14), the isotherms obtained intersect one another at the isopotential point, with coordinates E_{iso} and a_{iso}. Since variations in E° and solution activities with temperature are complex, three isotherms for temperatures T_1, T_2 and T_3 will not intersect at exactly the same point but will give three isopotential points. The isopotential point is thus dependent on the temperatures of the isotherms. If the isopotential point is shifted to $E_{iso} = 0$ by an auxiliary voltage applied from the instrument or by suitable choice of cell, then $a_{iso} = a_{zero}$ and standardization performed at, say, temperature T_1 can be used for temperature T_2, provided that the electrode slope is corrected between the standardization and the measurement. If a_{iso} is not identical with a_{zero}, then correct measurement can only be done at the standardization temperature. The a_{iso} value may differ substantially for an isothermal and a thermal cell* of the same composition [185]. As an example of the application of the isopotential technique, the control of fluorinated potable waters can be mentioned [183]. A symmetrical cell (i.e. the internal and external reference electrodes are identical) is employed, with an F^--ISE, the internal reference system of which has a fluoride concentration of 1 mg/l., which is the required F^--level, a KCl concentration of $1M$, and a silver chloride electrode. The external silver chloride reference electrode contains KCl at $1M$ concentration. It is evident from Fig. 2.15 that at around a fluoride concentration of 1 mg/l. (the isopotential point), i.e. in the region of interest, the cell voltage does not depend substantially on the temperature of the water tested. In principle, non-symmetrical cells can also be used for these purposes, but their selection may be more complicated [115].

The isopotential method has, of course, limited use, as it assumes that the dependence of the measured potential on temperature is the theoretical. There are deviations from the theoretical dependence, caused by various

---

* In an isothermal cell, both electrodes are at the same temperature, while in a thermal cell the reference electrode is placed outside the vessel and connected to the solution by a liquid bridge and generally has a different temperature from the ISE.

effects, one of the most important of which is the potential drift. This is a characteristic property of each electrode and changes with the electrode age and use. Not enough is known about the causes of the drift to allow *a priori* assessment of its approximate value. With some instruments the effect of temperature variations can be corrected manually by adjusting the temperature and isopotential controls; this can often be done only for a particular cell (mostly in pH measurement). On the other hand,

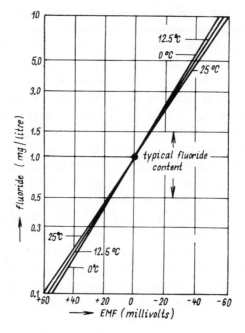

Fig. 2.15 The use of the isopotential technique for control of fluoridated waters. (Foxboro-Yoxall manufacturer's literature).

automatic correction by using an electrical offset [186] permits greater variability in E_{iso} and is thus applicable to various electrode systems.

Temperature coefficients of ISE's can be related to the standard hydrogen electrode (SHE) as a standard — these are isothermal temperature-coefficients $(dE_{SHE}/dT = 0)$ — or to an electrode maintained at a certain temperature — thermal or difference temperature-coefficients [185, 187]. When the temperature coefficient of an ISE is to be determined, thermal cells are usually employed (with the same or a different electrode as reference); an isothermal cell can also be used, provided that the temperature coefficient of the other electrode is known. For example, the SCE temperature coefficient can be determined from the cell

$$^-Pt \mid H_2 \mid H^+(a = 1) \parallel KCl \text{ (sat.)} \mid Hg_2Cl_2 \mid Hg^+$$

for which dE_c/dT is *ca.* -0.66 mV/deg (see Table 2.3). This is the isothermal coefficient for the SCE. Using the value $+0.88$ mV/deg given for $(dE/dT)_{th}$ of the SHE [187], the thermal value is

$$dE_c/dT = -0.66 = -(dE/dT)_{SHE} + (dE/dT)_{SCE} = -0.88 + x;$$
$$x = +0.22 \text{ mV/deg}$$

For $4M$ KCl and saturated silver chloride electrodes, a value of $+0.09$ mV/deg is given [149, 185]; however, the thermal coefficient increases with decreasing KCl concentration in both the calomel and the silver chloride electrode, in contrast to the isothermal coefficient [149]. So far, there is a lack of data concerning ISE temperature coefficients [185, 188, 189]; they are not quoted by the electrode manufacturers and may differ considerably for electrodes of different make, as they depend both on the membrane material and on the internal system.

REFERENCES

1. G. Johansson and K. Edström: *Talanta,* **19,** 1623 (1972).
2. W. Simon, W. Möller and R. Dohner: *US Patent* 3 647 666 (12. 6. 1969).
3. W. Möller: *Brit. Patent* 1 268 291 (12. 6. 1969).
4. J. Růžička, C. G. Lamm and J. C. Tjell: *Anal. Chim. Acta,* **62,** 15 (1972).
5. F. G. K. Baucke: *BRD Patent* 2 059 559 (3.12. 1970).
6. R. Bloch, M. Furmansky and S. Gassner: *BRD Patent* 2 109 664 (27. 2. 1970).
7. D. N. Gray and P. J. Breno: *US Patent* 3 806 440 (2. 4. 1973).
8. Beckman Instr., Inc.: *Brit. Patent* 872 486 (3. 3. 1958).
9. H. Kubota: *Anal. Chem.,* **42,** 1593 (1970).
10. A. E. Bottom and A. K. Covington: *J. Electroanal. Chem.,* **24,** 251 (1970).
11. J. H. Riseman and R. A. Wall: *US Patent* 3 306 837 (9. 5. 1963).
12. A. J. Petersen and G. Matsuyama: *US Patent* 3 649 506 (14. 10. 1969) and 3 718 569 (29. 11. 1971).
13. Carl-Zeiss Stiftung: *Brit. Patent* 1 324 839 (13. 8. 1970).
14. J. Veselý, O. J. Jensen and B. Nicolaisen: *Anal. Chim. Acta,* **62,** 1 (1972).
15. M. Koebel: *Anal. Chem.,* **46,** 1559 (1974).
16. M. Sato: *Electrochim. Acta,* **11,** 361 (1966).
17. R. P. Buck and V. R. Shepard, Jr.: *Anal. Chem.,* **46,** 2097 (1974).
18. J. Veselý and J. Jindra: *Proc. IMEKO Symp. Electrochem. Sens.* p. 69, Veszprém, Hungary (1968); *Chem. Abstr.,* **72,** 27534a (1970).
19. J. Körbl, E. Kraus, J. Martínek, J. Veselý and J. Jindra: *Czech Patent* 155 808 (17. 6. 1968); *Chem. Abstr.,* **83,** 52924 n (1975).

20. Perkin Elmer Corp.: *Brit. Patent* 1 240 028 (5. 8. 1968) and 1 298 719 (25. 4. 1969).
21. Radiometer a/s: *Brit. Patent* 1 269 019 (14. 7. 1969).
22. J. Růžička and C. G. Lamm: *Anal. Chim. Acta*, **53**, 206 (1971).
23. V. Majer, J. Veselý and K. Štulík: *J. Electroanal. Chem.*, **45**, 113 (1973).
24. J. Růžička and J. C. Tjell: *Anal. Chim. Acta*, **51**, 1 (1970).
25. J. P. Sapio, J. F. Colaruotolo and J. M. Bobbitt: *Anal. Chim. Acta*, **67**, 240 (1973).
26. Š. Mesarić and E. A. M. F. Dahmen: *Anal. Chim. Acta*, **64**, 431 (1973).
27. A. Hulanicki, R. Lewandowski and M. Maj: *Anal. Chim. Acta*, **69**, 409 (1974).
28. Hydronautics-Israel Ltd.: *Israel Patent* 35 473 (19. 10. 1970); *Chem. Abstr.*, **81**, 85719m (1974).
29. A. Ansadi and S. I. Epstein: *Anal. Chem.*, **45**, 595 (1973).
29a. H. Freiser: *Res./Dev.*, **27**, 28 (1976).
30. R. W. Cattrall and H. Freiser: *Anal. Chem.*, **43**, 1905 (1971).
31. H. Freiser, H. J. James, C. D. Carmack, B. M. Kneebone and R. W. Cattrall: *Brit. Patent* 1 375 785 (19. 1. 1972).
32. T. J. Rohm and G. G. Guilbault: *Anal. Chem.*, **46**, 590 (1974).
33. O. Ryba, E. Knižáková and J. Petránek: *Collection Czech. Chem. Commun.*, **38**, 497 (1973).
34. H. Hirata and K. Date: *Talanta*, **17**, 883 (1970).
35. T. Anfält and D. Jagner: *Anal. Chim. Acta*, **56**, 477 (1971).
35a. R. E. van de Leest, M. N. Bfekmans and L. Heijne: *BRD Patent* 2 600 846 (24. 1. 1975).
35b. R. W. Cattrall, D. M. Drew and I. C. Hamilton: *Anal. Chim. Acta*, **76**, 269 (1975).
36. A. Liberti: in *Ion-Selective Electrodes*, (E. Pungor and I. Buzás, eds.), p. 37, Akadémiai Budapest (1973).
37. S. L. Shiller and M. S. Frant: *US Patent* 3 442 782 (24. 5. 1966).
38. G. M. Farren and J. J. Staunton: *US Patent* 3 607 710 (14. 2. 1969).
39. W. T. Grubb: *BRD Patent* 2 158 243 (2. 12. 1970).
40. J. Veselý and K. Štulík: *Anal. Chim. Acta*, **73**, 157 (1974).
41. D. C. Stockbarger: *Disc. Faraday Soc.*, **5**, 294, 299 (1949).
42. V. G. Bamburov, N. V. Bausova, L. I. Manakova and A. P. Sivoplyac: *USSR Patent* 343 211 (8. 12. 1970); *Chem. Abstr.*, **77**, 159862u (1972).
43. A. M. G. MacDonald and K. Tóth: *Anal. Chim. Acta*, **41**, 99 (1968).
44. J. Veselý, J. Grégr and J. Jindra: *Czech. Patent* 143 144 (23. 6. 1969).
45. M. Sharp and G. Johansson: *Anal. Chim. Acta*, **54**, 13 (1971).
46. H. Hirata and K. Higashiyama: *Talanta*, **19**, 391 (1972).
47. J. D. Czaban and G. A. Rechnitz: *Anal. Chem.*, **45**, 471 (1973).
48. Orion Res., Inc.: *Brit. Patent* 1 150 698 (2. 10. 1967).
49. Orion Res., Inc.: *Brit. Patent* 1 222 476 (6. 9. 1968).
50. A. M. H. Weelink, G. W. S. van Osch and J. van Honwelingen: *US Patent* 3 824 170 (15. 1. 1973).
50a. I. Sekerka and J. F. Lechner: *Anal. Letters*, **9**, 1099 (1976).
51. J. Veselý and M. Lébl: *Czech. Patent* 151 612 (23. 3. 1970).
52. A. V. Gordievskii, V. S. Shterman, Yu. I. Urusov, N. I. Savvin, A. Ya. Syrchenkov and A. F. Zhukov: *USSR Patent* 397 832 (21. 2. 1972); *Chem. Abstr.* **81**, 20582s (1974).
53. M. S. Mohan and G. A. Rechnitz: *Anal. Chem.*, **45**, 1323 (1973).

54. J. C. van Loon: *Anal. Chim. Acta,* **54,** 23 (1971).
55. A. K. Covington: in *Ion-Selective Electrodes* (R. A. Durst, ed.), p. 89, NBS Special Publ. 314, Washington 1969.
56. E. Pungor: *Anal. Chem.* **39,** No. 13, 28A (1967).
57. E. Pungor, J. Havas, K. Tóth and L. Madarasz: *French Patent* 1 402 343 (22. 7. 1963).
58. A. M. Saunders: *US Patent* 3 709 811 (21. 9. 1970).
59. M. Mascini and A. Liberti: *Anal. Chim. Acta,* **47,** 339 (1969).
60. J. C. van Loon: *J. Environ. Anal. Chem.,* **3,** 53 (1973).
61. A. V. Gordievskii, N. I. Savvin, V. S. Shterman, A. Ya. Syrchenkov, A. F. Zhukov, A. S. Levin and Yu. E. Gotgel'f: *USSR Patent* 336 584 (25. 6. 1970); *Chem. Abstr.,* **80,** 152 532g (1974).
62. J. W. Ross, Jr.: *US Patent* 3 429 785 (17. 4. 1964); 3 445 365 (10. 8. 1965); 3 438 886 (14. 3. 1966).
63. J. W. Ross, Jr.: in *Ion-Selective Electrodes* (R. A. Durst, ed.), NBS Special Publ. 314, Washington 1969.
64. R. J. Settzo and W. M. Wise: *US Patent* 3 448 032 (3. 5. 1966).
65. H. Berge and P. Hartmann: *DDR Patent* 97 057 (18. 5. 1972).
65a. E. M. Skobets, L. I. Makovetskaya and Yu. P. Makovetskii: *Zh. Analit. Khim.,* **29,** 845 (1974).
66. R. E. Cosgrove, J. H. Krull and C. A. Mask: *US Patent* 3 715 297 (20. 7. 1970).
67. R. Bloch, D. Vofsi, O. Kedem and A. Katchalsy: *US Patent* 3 450 631 (13. 12. 1964).
68. N. Lakhsminarayanaiah: *Chem. Rev.,* **65,** 491 (1965).
69. O. Kedem, E. Loebel and M. Furmansky: *US Patent* 3 758 887 (4. 6. 1969).
69a. M. Perry, E. Loebel and R. Bloch: *J. Membr. Sci.,* **1,** 223 (1976).
70. U. Fiedler and J. Růžička: *Anal. Chim. Acta,* **67,** 179 (1973).
71. M. Mascini and F. Pallozzi: *Anal. Chim. Acta,* **73,** 375 (1974).
71a. U. Fiedler: *Anal. Chim. Acta,* **89,** 111 (1977).
72. G. Baum and M. Lynn: *Anal. Chim. Acta,* **65,** 393 (1973).
73. A. Shatkay: *Anal. Chem.,* **39,** 1056 (1967).
74. G. J. Moody, R. B. Oke and J. D. R. Thomas: *Analyst,* **95,** 910 (1970).
75. A. Craggs, G. J. Moody and J. D. R. Thomas: *J. Chem. Educ.,* **51,** 541 (1974).
75a. G. J. Moody and J. D. R. Thomas: in *Ion-Selective Electrodes* (E. Pungor, ed.) p. 41, Akadémiai Kiadó, Budapest (1977).
76. G. H. Griffiths, G. J. Moody and J. D. R. Thomas: *Analyst,* **97,** 420 (1972).
77. J. W. Ross , Jr. and M. S. Frant: *US Patent* 3 691 047 (8. 1. 1970).
78. L. W. Niedrach: *BRD Patent* 2 251 287 (21. 10. 1971).
79. Duke University: *Brit. Patent* 1 250 635 (20. 12. 1967).
80. Beckman Instr., Inc.: *Brit. Patent* 1 289 916 (29. 6. 1970).
81. H. Watanabe and J. E. Leonard: *US Patent* 3 635 212 (3.3.1969).
82. A. Strickler and C. H. Beebe: *US Patent* 3 649 505 (3. 3. 1969).
83. J. W. Ross, Jr., J. H. Riseman and J. A. Krueger: *IUPAC Intern. Symposium on Ion-Selective Electrodes,* Cardiff 1973, p. 473. Butterworth, London (1973).
84. J. Růžička and E. H. Hansen: *Anal. Chim. Acta,* **69,** 129 (1974).
85. E. H. Hansen and N. R. Larsen: *Anal. Chim. Acta,* **78,** 459 (1975).
86. C. B. Alcock (ed.): *Electromotive Force Measurements in High Temperature Systems,* Elsevier, New York (1968).
87. T. H. Etsell and S. N. Flengas: *Chem. Rev.,* **70,** 339 (1970).

88. W. A. Fischer: *Arch. Eisenhuettenw.*, **78**, 422 (1967).
89. Y. Matsushita and K. Goto: *Thermodyn. Proc. Symp.*, **1**, 111 (1965).
90. G. G. Guilbault: *Enzyme Electrodes*, in *Practical Uses of Immobilized Enzymes*, H. Weltall (ed.), Dekker, New York (1974).
91. R. A. Llenado: *Thesis*, Univ. Microfilms, Ann Arbor, Mich., Order No. 74-4421 (1973).
92. C. Tran-Minh and G. Broun: *Anal. Chem.*, **47**, 1359 (1975).
92a. G. Johansson and L. Ögren: in *Ion-Selective Electrodes* (E. Pungor, ed.), p. 93, Akadémiai Kiadó, Budapest (1977).
93. G. G. Guilbault and M. Tarp: *Anal. Chim. Acta*, **73**, 355 (1974).
94. G. G. Guilbault and W. Stokbro: *Anal. Chim. Acta*, **76**, 237 (1975).
95. R. A. Durst and J. K. Taylor: *Anal. Chem.*, **39**, 1483 (1967).
96. R. A. Llenado and G. A. Rechnitz: *Anal. Chem.*, **43**, 1457 (1971).
97. J. H. Riseman and S. L. Shiller: *US Patent* 3 492 216 (9. 3. 1967).
97a. W. Simon, E. Pretsch, D. Ammann, W. E. Morf, M. Güggi, R. Bissig and M. Kessler: *Pure Appl. Chem.*, **44**, 613 (1975).
97b. H. J. Berman and N. C. Hébert: *Ion-Selective Microelectrodes*, Plenum Press, New York (1974).
98. M. Lavallé, O. F. Scanne and N. C. Hébert (eds.): *Glass Microelectrodes*, Wiley, New York (1969).
99. J. A. M. Hinke: in *Glass Electrodes for Hydrogen and Other Cations, Principles and Practice*, G. Eisenman (ed.), p. 464. Dekker, New York (1967).
100. R. N. Khuri: in ref. 99, p. 478.
100a. D. W. Luebbers, T. Shigemitsu and H. Baumgaertl: *Naturwiss.*, **63**, 40 (1976).
101. J. A. M. Hinke: *Nature*, **184**, 1257 (1959).
102. J. L. Walker: *Anal. Chem.*, **43**, No. 3, 89A (1971).
103. J. L. Walker: in *Ion-Specific Microelectrodes*, (N. C. Hébert and R. N. Khuri, eds.), Dekker, New York (1973).
104. R. A. Steinhardt: *US Patent* 3 743 591 (3. 7. 1973).
105. F. Vyskočil and N. Kříž: *Pflügers Arch.*, **337**, 265 (1972).
106. P. Hník, N. Kříž, F. Vyskočil, V. Smieško, J. Mejsnar, E. Ujec and M. Holas: *Pflügers Arch.*, **338**, 117 (1973).
107. P. Bergveld, *IEEE Trans.*, BME-19, 342 (1972).
108. T. Matsuo and K. D. Wise, *IEEE Trans.*, BME-21, 485 (1974).
109. J. N. Zemel: *Anal. Chem.*, **47**, 255A (1975).
110. S. D. Moss, J. Janata and C. C. Johnson: *Anal. Chem.*, **47**, 2238 (1975).
110a. *Proceedings of Workshop on Theory, Design and Biomedical Applications of Solid State Chemical Sensors*, (P. W. Cheung. ed.). CRC Press, Cleveland (1978).
111. D. W. Hoole, G. L. Klein and T. R. Vivian: *US Patent* 3 575 834 (12. 4. 1968).
112. H. Makabe: *US Patent* 3 666 651 (2. 7. 1968).
112a. J. Mertens, P. van den Winkel and D. L. Massart: *Anal. Chem.*, **48**, 272 (1976).
113. D. J. G. Ives and G. J. Janz: *Reference Electrodes*, Academic Press, New York (1961).
114. A. K. Covington: in *Ion-Selective Electrodes* (R. A. Durst, ed.), p. 107, NBS Special Publ. 314, Washington (1969).
115. R. A. Bates: *Determination of pH*, Wiley, New York (1973).
116. M. J. Barbier and J. J. Rameau: *Bull. Soc. Chim. France*, 126 8 (1973).
117. G. J. Janz and H. Taniguchi: *Chem. Rev.*, **53**, 397 (1953).
118. G. J. Janz and D. J. G. Ives: *Ann. N. Y. Acad. Sci.*, **148**, 210 (1968).

119. A. S. Brown: *J. Am. Chem. Soc.*, **56**, 646 (1934).
120. C. K. Rull and V. K. LaMer: *J. Am. Chem. Soc.*, **58**, 2339 (1936).
121. D. T. Ferrell, Jr., I. Blackburn and W. C. Vosburgh: *J. Am. Chem. Soc.*, **70**, 3812 (1948).
122. H. D. Portnoy: in ref. 99, p. 257.
122a. C. Harzdorf: Private communication.
123. G. J. Janz: ref. 113, p. 179.
124. G. J. Moody, R. B. Oke and J. D. R. Thomas: *Analyst*, **94**, 803 (1969).
125. J. V. Dobson, R. E. Firman and H. R. Thirsk: *Electrochim. Acta*, **16**, 793 (1971).
126. H. K. Fricke: in ref. 114.
127. Radiometer a/s: *Brit. Patent* 1 281 116 (28. 6. 1969).
128. R. G. Bates and V. E. Bower: *J. Res. Natl. Bur. Stand.*, **53**, 283 (1954).
129. R. S. Greeley, W. T. Smith, Jr., R. W. Stoughton and M. H. Lietzke: *J. Phys. Chem.*, **64**, 652 (1960).
130. F. Štráfelda: *Chem. Listy*, **50**, 190 (1956).
131. R. A. Robinson and R. H. Stokes: *Electrolyte Solutions*, p. 463, Butterworths, London (1959).
132. H. B. Hetzer, R. A. Robinson and R. G. Bates: *J. Phys. Chem.*, **66**, 1423 (1962).
133. M. B. Towns, R. S. Greeley and M. H. Lietzke: *J. Phys. Chem.*, **64**, 1861 (1960).
134. H. B. Hetzer, R. A. Robinson and R. G. Bates: *J. Phys. Chem.*, **68**, 1929 (1964).
134a. F. G. K. Baucke: *J. Electroanal. Chem.*, **67**, 277 (1976).
135. H. F. Gibbard, Jr.: *J. Electrochem. Soc.*, **120**, 624 (1973).
136. J. Číhalík: *Potenciometrie*, NČSAV, Prague (1961).
137. D. J. Alner, J. J. Greczek and A. G. Smeeth: *J. Chem. Soc. A*, 1205 (1967).
138. G. J. Hills and D. J. G. Ives: *J. Chem. Soc.*, 311, 318 (1951).
139. S. N. Das and D. J. G. Ives: *J. Chem. Soc.*, 1619 (1962).
140. D. J. G. Ives and D. Prasad: *J. Chem. Soc. B*, 1649 (1970).
141. N. Linnet: *pH Measurements in Theory and Practice*, Radiometer, Copenhagen (1970).
142. F. Štráfelda and B. Polej: *Chem. Průmysl*, **7**, 240 (1957).
143. A. K. Covington, J. V. Dobson and Lord Wynne-Jones: *Electrochim. Acta*, **12**, 525 (1967).
144. S. R. Gupta, G. J. Hills and D. J. G. Ives: *Trans. Faraday Soc.*, **59**, 1874 (1963).
145. F. G. K. Baucke: *J. Electroanal. Chem.*, **33**, 135 (1971).
146. F. G. K. Baucke: *Chem. Ing. Tech.*, **46**, 71 (1974).
147. G. J. Hills and D. J. G. Ives: ref. 113, Chapter 3.
148. A. K. Covington, J. V. Dobson and Lord Wynne-Jones: *Trans. Faraday Soc.*, **61**, 2050 (1965).
149. G. Mattock: *pH Measurement and Titration*, Heywood, London (1961).
150. B. Gutbezahl and E. Grunwald: *J. Am. Chem. Soc.*, **75**, 565 (1953).
151. E. A. Guggenheim: *J. Am. Chem. Soc.*, **52**, 1315 (1930).
152. W. H. Smyrl and J. Newmann: *J. Phys. Chem.*, **72**, 4660 (1968).
153. G. A. Perley: *Trans. Electrochem. Soc.*, **92**, 497 (1947).
154. G. Mattock, in: *Advances in Analytical Chemistry and Instrumentation* (C. N. Reilley, ed.), Vol. 2, Wiley, New York (1963).
155. W. N. Carson, Jr., C. E. Michelson and K. Koyama: *Anal. Chem.*, **27**, 472 (1955).
156. L. B. Leonard and H. Watanabe: *US Patent* 3 264 205 (30. 7. 1963).
157. M. L. Deushane and D. A. Rohrer: *US Patent* 3 741 884 (4. 5. 1972).
158. E. C. ZoBell and S. C. Rittenberg: *Science*, **80**, 502 (1937).

159. J. T. Taylor: *US Patent* 3 528 903 (24. 1. 1968).
160. F. Čůta and F. Straka: *Chem. Listy*, **67**, 861 (1973).
161. Beckman Instr., Inc.: *Brit. Patent* 1 134 140 (18. 6. 1965).
162. Beckman Instr., Inc.: *Brit. Patent* 1 337 829 (21. 8. 1970).
163. G. Brunisholz: *Anal. Chim. Acta*, **10**, 470 (1954).
164. Okura Denki Kabushiki Kaisha: *Brit. Patent* 1 070 674 (10. 3. 1964).
165. L. Jenšovský: *Chem. Listy*, **44**, 75 (1950).
166. J. N. Brønsted: *Trans. Faraday Soc.*, **23**, 430 (1927).
167. M. Spiro: *Electrochim. Acta*, **11**, 569 (1966).
168. P. Henderson: *Z. Physik. Chem.*, **59**, 118 (1907); **63**, 325 (1908).
169. E. W. Baumann: *J. Electroanal. Chem.*, **34**, 238 (1972).
170. A. Shatkay and A. Lerman: *Anal. Chem.*, **41**, 514 (1969).
171. P. A. Rock: *Electrochim. Acta*, **12**, 1531 (1967).
172. Yu. M. Chernoberezhskii, S. N. Zulkova, S. D. Usanova and L. V. Afanasieva: *Koll. Zh.*, **27**, 780 (1965).
173. H. Jenny, T. R. Nilsen, N. T. Coleman and D. E. Williams: *Science*, **112**, 164 (1950).
174. V. K. LaMer: *J. Phys. Chem.*, **66**, 973 (1962).
175. H. Malmwig: *GIT Fachz. Lab.*, **19**, 402 (1975); *Chem. Abstr.*, **84**, 11746h (1976).
176. A. F. Isbell, Jr., R. L. Pecsok, R. H. Davies and J. H. Purnell: *Anal. Chem.*, **45**, 2363 (1973).
177. I. Sekerka and J. F. Lechner: *Intern. J. Environ. Anal. Chem.*, **2**, 313 (1973).
178. I. Sekerka and J. F. Lechner: *Talanta*, **22**, 459 (1975).
179. Orion Res., Inc.: *Newsletter*, **6**, No. 2 (1974).
180. E. L. Eckfeld and G. Perley: *J. Electrochem. Soc.*, **98**, 37 (1951).
181. W. M. Krebs: *Anal. Chem.*, **44**, 187 (1972).
182. T. S. Light: in *Ion-Selective Electrodes*, (R. A. Durst, ed.), NBS Special Publ. 314, Washington (1969).
183. R. H. Babcock and K. A. Johnson: *J. Am. Water Works Assoc.*, **60**, 953 (1968).
184. J. Jackson: *Chem. Ind. London*, 7 (1948).
185. L. E. Negus and T. S. Light: *Instrum. Technol.*, **19**, 23 (1972).
186. A. J. Williams, Jr.: *US Patent* 2 674 719.
187. A. J. deBethune, T. S. Light and N. Swendeman: *J. Electrochem. Soc.*, **106**, 616 (1959).
188. I. Sekerka and J. F. Lechner: *Tech. Bull.* No. 72, Canada Centre for Inland Waters, Burlington, Ontario (1973).
189. E. Lindner, K. Tóth and E. Pungor: in *Ion-Selective Electrodes* (E. Pungor and I. Buzás, eds.) p. 205, Akadémiai Kiadó, Budapest (1973).

Chapter 3

EXPERIMENTAL TECHNIQUES

Measurement with ISE's can be divided into three principal steps.

(1) Conversion of the sample into a form suitable for potentiometric measurement with the particular electrode.

(2) Selection of the optimum measuring techniques from the point of view of the selectivity, sensitivity, accuracy, precision and kind of information required.

(3) Selection of the optimum procedure for the evaluation of the measurements.

3.1 PREPARATION OF THE SAMPLE FOR THE MEASUREMENT

3.1.1 Sample decomposition

The measurement can be performed only on a suitable solution. If the sample is not sufficiently soluble in water (or possibly in some other polar medium), it must be decomposed; the decomposition procedure must be chosen individually, depending on the character of the material and the species to be determined. The chief requirements for the decomposition procedure are (a) the species to be determined must be quantitatively transferred into the solution, (b) the amounts of interfering substances and reagents transferred into the solution must be as low as possible, (c) the procedure should be simple and rapid.

The decomposition techniques available fall into several groups.

(1) Decomposition with mineral acids

(2) Fusion

(3) Sintering

(4) Leaching or extraction

(5) Distillation

(6) Mineralization

(7) Sorption and absorption

The first three groups find use especially in the decomposition of mineral raw materials [1]. Dilute or concentrated hydrochloric, nitric, perchloric and sulphuric acids are mostly used, sometimes along with hydrofluoric acid when silicates are to be decomposed. The insoluble residues are then fused with a suitable substance. Decomposition with acids is especially suitable for determinations with cation-sensitive electrodes. Decomposition by fusion is more suitable when anions are to be determined in materials which are poorly soluble in acids. As reagents, alkali metal carbonates or hydroxides and their mixtures are most often used [2, 3]. The main disadvantages of these procedures for measurement with ISE's are that the reagents employed often contain impurities, the ionic strength of the solution is very high and it may be difficult to adjust the pH to the value required.

Therefore, sintering is usually more convenient. A mixture of an alkali metal carbonate and a bivalent metal oxide is most often used. After sintering, the material is leached with water, giving solutions free from most interfering cations, which remain in the insoluble residue as the carbonates. These procedures are mostly applied when anions are to be determined; sodium carbonate mixed with zinc oxide and sometimes with sodium nitrate added as an oxidant [4] has proved advantageous. Leaching or extraction of samples with water, dilute mineral acids and solutions of salts, and sometimes with organic solvents, is used e.g. in pedology, agricultural sciences, in the building materials industry and in geological prospecting. When organic solvents are employed, it is usually necessary to back-extract the species to be determined into an aqueous phase [5]; sometimes, however, the measurement can be directly performed on the organic phase [6]. Leaching and extraction procedures are advantageous in that the effect of interfering species is suppressed and relatively small amounts of impurities are introduced into the system.

Distillation is chiefly used when the species of interest is to be separated from interfering components, with simultaneous preconcentration. The technique can be used when the species to be determined forms a volatile compound; the measurement is then performed directly on the distillate [7]. Pyrolytic or pyrohydrolytic [8] methods are similar. The species determined is converted into a volatile compound at a high temperature in the presence of a catalyst, and of steam in pyrohydrolysis. This compound is then absorbed in water or a suitable absorbent, or is transferred directly into the aqueous condensate. These methods are very suitable for measurement with ISE's [9], as the resultant solutions are practically free from interfering components. A serious disadvantage is their tediousness.

Mineralization finds important use in the decomposition of organic materials. The samples are either combusted in air or oxygen, or mineralized by heating with strong oxidants, such as nitric and perchloric acids or hydrogen peroxide, often mixed with sulphuric acid. The residue is then dissolved in water or decomposed by some of the techniques above. Organic materials can also be decomposed by ignition with sodium carbonate, hydroxide or peroxide [10].

The sorption and absorption methods are primarily separational processes; however, some of them can also be used for decomposition of various materials. For example, ion-exchangers can be used for decomposition of fluorophosphates, with simultaneous separation of the components of the mixture [11]. These methods are often applied in the analysis of gases and aerosols, the species to be determined being collected in a suitable absorption solution or on a membrane filter impregnated with an absorbent [12]. The latter technique is to be preferred, as the absorbent volume is small and consequently the determination is more sensitive.

Application of ISE's in analysis of waters and some biological fluids (such as serum or urine) is especially important, as these contain ions the concentrations or activities of which can usually be measured without preliminary operations. The samples should therefore be collected in tightly stoppered plastic bottles and analysed in the field or quickly transported to the laboratory. If it is impossible to analyse the samples immediately, they must be suitably conserved. Water samples can be stored for some time at low temperature (down to about 5 °C) in a cooling-box. Only for the determination of nitrates and nitrites is it recommended that the samples be preserved by adding *ca.* 1 ml of chloroform per litre of water; a special preservative solution is added when sulphides are to be determined [13]. Sometimes, especially when cations are to be determined, water can be preserved by saturation with carbon dioxide. Traditional methods for the preservation of water samples, e.g. by acidification with a mineral acid, are rarely suitable for measurements with ISE's, since undesirable ions are introduced into the system and neutralization leads to an increase in the ionic strength. If blood samples for the determination of calcium ions must be stored, 5–10 units of heparin per ml of blood (not more) are introduced into the syringe to prevent clotting. It is not advisable to store the sample at room temperature; refrigeration is preferable to freezing. If possible the original sample pH should be restored before the measurement by using a CO_2–O_2 mixture, in order to avoid changes in the content of calcium ions due to pH variations (inverse proportionality exists between the ionic calcium content and the pH). The sample must be collected under anaerobic conditions to prevent loss of CO_2 [14, 15].

3.1.2 Preparation of the sample solution for the measurement

As follows from Chapters 1 and 2, the main factors affecting the potential difference of a cell containing an ISE are, in addition to the activity of the species to be determined, the kind and activities of interfering species, the pH, the ionic strength and the temperature. Therefore, the effect of interferents must be suppressed and the other factors kept constant. To maintain a constant temperature, the solution must be kept in a thermostat for very precise measurements and otherwise kept at a temperature constant within $\pm 0.5\ °C$ or better. It is usually better to measure at higher temperature, about 25 °C, where the response time and the precision of the electrodes are best.

The response of all ISE's depends on the pH; for some electrodes there are certain optimum pH ranges, others require accurate adjustment to a certain pH. It is almost always true that electrodes that can tolerate a certain variation in the pH exhibit smaller sensitivity towards pH changes when the concentration of the species sensed increases. The required pH value can be adjusted in an unbuffered solution, but it is more reliable to use suitable buffers if possible. It is advantageous to prepare buffers in such a way that they perform a multiple function, namely, maintain a constant ionic strength and mask some interfering species in addition to their pH-buffering action. In order to keep the activity coefficients and the liquid-junction potentials reasonably constant, the ionic strength of the solution should be constant, within the range $ca.$ $0.1-2M$. A typical example of a buffer which adjusts the pH and the ionic strength and masks some components of the system (TISAB — Total Ionic Strength Adjustment Buffer) [16], is that containing an acetate pH-buffer, nitrate for the ionic strength adjustment and citrate for masking some ions, e.g. iron and aluminium. The complexing components of buffers can, however, unfavourably affect the electrode function, either by affecting the activity of the species sensed or by interacting with the electrode membrane. The buffers used usually have ionic strengths $> 2M$ and are mixed with the sample solution in a 1 : 1 ratio. Sometimes it is better to add less of a more concentrated buffer in order not to dilute the sample solution too much. When the sample itself contains a large amount of salts (concentration more than about $2M$), the electrode must be calibrated with simulated standards containing the same amount of salts. When the activity is to be determined, most of the operations above are unnecessary. Nevertheless, the measurement should be performed under the same conditions as those existing during sampling.

3.2 DISCONTINUOUS MEASUREMENTS

In discontinuous measurements the sample solutions are compared with standards by use of a calibration curve or ion-meters (calibrated with two or three standards) or addition methods are employed; the sample can also be titrated. The first two techniques (direct potentiometry) are simple, rapid procedures which, however, exhibit poorer precision than potentiometric titrations. The precision of direct potentiometry depends primarily on the value of the slope of the E vs. a plot (and therefore decreases with decreasing slope) and also on the effectiveness of suppression of the effect of temperature variations, interfering components, the liquid-junction potential and changes in the electrode characteristics during the measurement.

Generally, more reproducible results and more rapid responses are obtained in vigorously stirred solutions, where better agreement of the E vs. a dependence with theory is also achieved. Mechanical (vertical) stirrers are more suitable than magnetic stirrers, as they provide more regular and intense solution movement. Moreover, some magnetic stirrers cause heating of the measured solution; this can be prevented by placing a thin plate of a heat-insulating material underneath the vessel. The stirring should not be so violent that air bubbles are drawn into the solution. A rotating ISE has also been constructed, ensuring intense regular movement of the measured solution relative to the ISE [17].

During the measurements, two kinds of interference can occur: the electrode interference, expressed by the selectivity coefficient for the particular electrode and interfering species (see Chapter 1), and chemical interferences caused by interactions of components of the solution with the electrode membrane or with the species sensed by the electrode. The products of these chemical reactions or other components of the solution (e.g. species of high molecular weight) can be adsorbed on the electrode surface and cause deterioration in its response; the electrode must then be regenerated. If the poisoning of the electrode is reversible, chemical regeneration is usually sufficient; the electrode is immersed in a suitable solution in which the deposit on the membrane is desorbed or dissolved. When the poisoning is irreversible (e.g. the effect of sulphide on silver halide electrodes or that of surfactants present in biological samples), the membrane surface must be renewed mechanically by polishing with metallographic papers or with a suitable abrasive paste; in liquid-state electrodes the liquid ion-exchanger must be replaced. Then the electrode is immersed in a solution of the species sensed, at a concentration of about $10^{-3}M$. Generally, poisoning is more serious when the concentration of the interfering species increases and its action is more prolonged [18]. Poisoning can sometimes be prevented by suitable preparation of standard solutions (e.g. addition of trypsin and triethanol-amine in order to dissolve proteins [19]). The "memory" effect, observed particularly with solid-state electrodes, is similar to poisoning. This is apparently also caused by adsorption and desorption, but this time of the ion sensed by the electrode and hence is a sort of self-poisoning of the electrode. This effect is manifested during measurements in more dilute sample solutions, as a temporary positive error in the electrode potential after use of the electrode for measurements on solutions of the same species at concentrations above about $10^{-3}M$. The longer the electrode is exposed to high concentration of the ion, the greater the memory effect up to a certain limit. Rinsing and wiping the electrode before it is transferred to the dilute solution does not help, but the disappearance of the effect is hastened by vigorous stirring of the dilute solution. On the other hand, it is not advisable to leave the electrode too long in very dilute solutions (10^{-5}–$10^{-7}M$), as the reverse effect occurs when the electrode is used for measurements in more concentrated solutions. This latter effect is pronounced with liquid ion-exchanger electrodes, but it has also been observed with some solid-state electrodes and is probably caused by removal of the determinand ion from the membrane. Immersion of the electrode in a solution that is about $10^{-3}M$ in the ion measured, and vigorous stirring, again hasten electrode regeneration. Adsorption and desorption of the species from the membrane surface is indicated by the stirring effect, i.e. $\Delta E = E_{stirred} - E_{unstirred}$ when the activity of the species sensed and the ionic strength are low. If the activity reading is higher in an unstirred solution than during stirring, desorption has occurred, and if it is lower, adsorption [20]. The effect of stirring also depends on the method of membrane preparation. The sudden appearance of a large stirring effect may also reflect undesirable changes in the liquid-junction potential.

"Memory" effects can seriously affect the reliability of measurements, if the electrode function is not regularly checked, preferably by re-calibration.

The question of preconditioning electrodes is an important one. The manufacturers usually state, especially with solid-state electrodes, that no preliminary conditioning is necessary. However, practice has shown that the response of non-conditioned electrodes is unreliable, sluggish and poorly reproducible; the potentials exhibit drift, especially at very low and very high concentrations. Lowered sensitivity and deviations from Nernstian response can also be observed. Substantial improvement is achieved if the electrode is immersed for at least one hour in a *ca.* $10^{-3}M$ solution of the determinand species before the measurement. Then the electrode is conditioned by repeated measurements at various concentrations of the species sensed, avoiding very low and very high concentrations in order to avoid memory effects. Glass and heterogeneous electrodes require special preconditioning procedures.

The response rate varies according to the electrode type. Generally, solid-state electrodes respond substantially faster than the liquid ion-exchanger types. At concentrations above $10^{-4}M$ the response times typically amount to several seconds or tens of seconds. With decreasing concentrations the response time increases and may be several minutes or tens of minutes. If potential stabilization takes more than about 5 minutes, the reading is usually taken after a certain constant time interval, even if the value still changes. The response is hastened by increasing the temperature and by stirring the solution. The response time is also affected by the state of the membrane, the solution composition and the quality of the reference electrode.

The reproducibility of the measured potential depends on practically the same factors as the response rate. When optimum conditions are selected, the confidence limits (95 % probability) are typically around 0.5 pX for concentrations bordering on the detection limit of the electrode and about 0.02 pX for concentrations higher than *ca.* $10^{-4}M$. At concentrations higher than about $5 \times 10^{-2}M$ this value is about 0.05–0.1 pX.

The principles of correct measuring technique for concentration determinations in direct potentiometry with ISE's exhibiting Nernstian response and with good-quality instrumentation can be summarized as follows.

(1) The electrode is soaked in *ca.* $10^{-3}M$ solution of the species sensed, for at least one hour before the measurement.

(2) In all test and standard solutions a constant ionic strength is maintained by adding a suitable indifferent electrolyte or buffer.

(3) In all solutions measured a constant pH is maintained, or at least it is kept within a recommended interval.

(4) All solutions measured are maintained at a temperature constant within ± 0.5 °C.

(5) The volume of the solutions is kept constant.

(6) The electrode is always immersed to the same depth, in solutions placed in plastic beakers.

(7) The solutions are stirred vigorously at a constant speed, preferably with a vertical stirrer, the stirrer being kept in the same position with respect to the electrode and the vessel walls.

(8) The electrode is preconditioned by repeated measurements on standard solutions of various concentrations, until the electrode begins to respond reproducibly.

(9) The electrode is calibrated with standard solutions of the ion measured over about three decades of concentration. Simulated standards are used wherever possible and the electrode calibration is repeated regularly.

(10) Interfering species are masked with suitable complexing agents.

Only points 1, 4, 6, 7 and 8 are important when the activities are to be determined.

3.2.1 Calibration curves and simulated standards

When a commercial ion-meter is employed (see Chapter 2), the electrode is calibrated with one or, preferably, two standard solutions and then the approximate concentration values are read directly on the display, provided, of course, that the slope is constant over the given concentration range. Otherwise, an empirical calibration curve must be constructed, in terms of either the activities or the concentrations. For construction of an activity calibration curve the appropriate individual activity coefficients must be known (see Table B in the Appendix). However, for analytical practice the measurement of the overall concentration of the particular species is of greater importance; measurements of activities find their principal use in studies of equilibrium reactions and in systems where the ratio of the free and complexed species is important (e.g. calcium in blood).

A calibration curve is constructed by plotting the electrode potential against the negative logarithm of the suitably expressed concentration of the species. A series of standard solutions with a composition as close as possible to that of the sample is employed and the conditions are maintained identical with those used for the measurement on the sample. The best results are obtained with simulated standards, as all effects of the other components of the sample solution are then included in the calibration curve.

The standard addition method can also be used for the construction of the calibration curve, using the Gran method for evaluating the blank value (see Sections 3.2.2 and 3.2.4) or the species sensed can sometimes be generated coulometrically, or the test solution can be progressively diluted [21a].

The calibration curve method can also be used for the determination of species that are not sensed by the electrode, by use of an auxiliary reaction; for example, Al^{3+} can be determined by using F^- as the auxiliary ion or Ni^{2+} with CN^- as the auxiliary ion (see e.g. [21]).

Determinations with gas-sensing and enzyme electrodes are also indirect techniques. Modified methods are also used for indirect determination of the concentration of a catalyst in a suitable reaction by monitoring time-variations in the potential of an ISE sensitive to an ion participating in the reaction (see e.g. the determination of the rhodanese enzyme [22]), or by using calibration curves constructed by plotting the slope of the linear part of the E–t dependence against the catalyst concentration (e.g. the determination of Mo and W down to a concentration of 0.004 ppm [23]).

The concentration calibration curves are usually linear within a concentration range of ca. 10^{-2}–10^{-4} or $10^{-5}M$, with a slope more or less close to the theoretical Nernstian value. Curvature occurs at higher and especially lower concentrations and eventually the electrode does not respond to changes in the concentration at all. The curvature at high concentrations is caused by saturation of the membrane and at low concentrations by its dissolution or self-poisoning. If the electrode is carefully calibrated and the experimental conditions selected correctly and maintained constant, relatively satisfactory and reproducible results can be obtained even where curvature of the calibration curve occurs; the errors usually do not exceed 20 % when the concentration is above $5 \times 10^{-5}M$ or, with some electrodes, down to $5 \times 10^{-6}M$. Determinations of even lower concentrations require favourable conditions (e.g. low interferent contents, rapid ISE response etc.) and/or special procedures and have not yet found broader use in the analysis of natural materials. For example, very low concentrations (10^{-7}–$10^{-8}M$) can be determined semiquantitatively by measuring the potential alternately in the sample and in standards and assessing whether the sample concentration is higher or lower than that of the standard from the direction of the change in the unstabilized potential [24]. The method is very rapid and useful even if the electrode potential exhibits slight drift, which is common at low concentrations.

In contrast to concentrations, extremely low activities can sometimes be measured (down to about $10^{-20}M$), provided, of course, that the overall (analytical) concentration of the species (i.e. the sum of the concentrations of the free species and of all its forms bound in complexes and precipitates) is higher than *ca.* $10^{-6}M$. Fluctuations in very low activities due to sorption, hydrolysis, etc., are controlled by a suitable chemical equilibrium (see Section 1.3). This activity range ($a < ca.$ $10^{-8}M$) can be analytically used only in indirect determinations, but it is important in chemical equilibrium studies.

3.2.2 Determination of overall concentration by addition techniques

The concentration of a species can be successfully determined by using a calibration curve only when the amount of the species bound in complexes is negligible. On the other hand, addition techniques permit the determination of the overall concentration of species, even if a considerable part of the species is complexed. Another advantage of addition methods is the relative rapidity and simplicity of the procedure; moreover, the electrode need not always be calibrated. When addition methods are used, the concentration of the species to be determined is varied and the unknown concentration is determined from the difference of the potentials before and after the addition.

If a standard solution of the species to be determined is added, the procedure is called the known-addition method (KAM). When a single addition is made, the concentration can be determined only when the electrode response slope, S, is known. When several additions are made, the number of equations available increases, thus permitting the determination even if the slope is not known beforehand, and hence the electrode calibration is unnecessary; however, the precision is poorer. A modification of the KAM is the analyte-addition method, which is especially suitable for measurements on samples available only in small quantities, as the sample is added to the standard solution.

These methods can be operated in the reverse sense and the activity of the species to be determined decreased by the addition of a suitable solution. These are the known-subtraction (KSM) and analyte-subtraction (ASM) methods. For decreasing the concentration of the species determined, suitable chemical reactions that proceed stoichiometrically and sufficiently rapidly can be employed (complexation, precipitation or redox reactions). In the ASM some auxiliary ion, sensed by the electrode, is removed by the addition of the sample. The latter method is thus suitable for the determination of species for which no electrode is available.

The addition methods can be applied when several conditions are met, namely, that the individual activity coefficients of the substance to be determined, the fraction of the substance complexed or precipitated, the electrode response slope and the liquid-junction potential all remain constant during the additions and that there are no interferences. To keep the activity coefficient of the species constant, a sufficient and constant excess of an indifferent electrolyte must be present; then relative constancy of the liquid-junction potential is also achieved.

The analytical concentration of the species determined is changed by the addition of a standard solution, but the fraction of it bound in complexes or precipitates may still remain the same. For example, when fluoride is to be determined in an acidic medium, some undissociated hydrogen fluoride is present:

$$K_1 = \frac{[HF]}{[H^+][F^-]} \; ; \quad [H^+] . K_1 = \frac{[HF]}{[F^-]} = \alpha \qquad (3.1)$$

but α depends only on $[H^+]$ and not on $[F^-]_{tot}$; therefore in the presence of a sufficient excess of hydrogen ions, the α value is virtually not affected by the addition. However, care must be exercised when the concentrations of the species to be determined and the complexing or precipitating agent are similar, e.g. in the vicinity of the equivalence point of a titration, when even a small change in the concentration may lead to gross changes in α. It is then necessary to add a sufficient amount of an agent forming a more stable complex or precipitate with the species determined than that already present, or to liberate the species to be determined by a suitable method (e.g. by addition of TISAB).

The electrode response slope, S, exhibits greater stability with time than the absolute potential values do. However, it often depends on the activity of the species sensed (see the curvature of the calibration curves — Section 3.2.1), especially near the detection limit of the electrode. When the activity range used is narrow, this variation in the slope can be neglected, but the error of the determination increases, as the addition of a small amount of the substance leads to only a small change in the potential. The slope further depends on the temperature, but in addition methods the difference in the temperatures of the sample and the solution added is not critical, since usually a very small volume is added. Moreover, the problem of varying temperature can be elegantly circumvented, e.g. when measuring in the field, by using double or multiple addition methods, where the slope need not be known.

Interfering substances must generally be absent if the determination is to be successful. There are some ways of dealing with interferences in addition techniques, but they are very limited. When using the Gran graphical linearization method (see Section 3.2.4) it is possible to perform the determination in the presence of a constant amount of interfering species (e.g. after chemical treatment of the sample). It is also possible to utilize the fact that the effect of some interfering species rapidly decreases with dilution of the sample solution. In this way the interference of Al^{3+} in the determination of fluoride in aluminium silicate glasses was suppressed [25] and the interference of univalent cations in the response of electrodes sensitive to bivalent cations might also be decreased similarly. For example, the serious interference of Na^+ in determinations with the Ca^{2+}-electrode decreases rapidly on dilution of the sample, as the electrode potential [see Eq. (1.32)] is proportional to the square of the sodium activity and to the first power of the calcium activity. Of course, the applicability of this method is limited by the detection limit of the electrode.

3.2.2.1 KNOWN-ADDITION AND KNOWN-SUBTRACTION METHODS

The simplest method is based on two potential readings, before (E_1) and after (E_2) the addition of a volume V_a of a standard solution to volume V_0 of the sample:

$$E_1 = E^{0'} \pm \frac{RT}{nF} \ln \frac{\gamma_1}{\alpha_1} C_x \qquad (3.2)$$

$$E_2 = E^{0'} \pm \frac{RT}{nF} \ln \frac{\gamma_2}{\alpha_2} \left(\frac{C_x V_0 \pm C_s V_a}{V_0 + V_a} \right) \qquad (3.3)$$

where + in the numerator of the logarithmic term refers to the KAM and − to the KSM. C_s denotes either the concentration of the standard solution or the known concentration decrease caused by a stoichiometric reaction in the KSM, and C_x is the concentration of the determinand in the sample.

The potential change is

$$\Delta E = E_2 - E_1 = \pm \frac{RT}{nF} \ln \frac{\gamma_2 \alpha_1}{\gamma_1 \alpha_2} \left[\frac{C_x V_0 \pm C_s V_a}{(V_0 + V_a) C_x} \right] \qquad (3.4)$$

If the conditions above are met, i.e. $\gamma_1 = \gamma_2$ and $\alpha_1 = \alpha_2$, and S is substituted for $\pm[(RT)/(nF)]\ln 10$, then

$$\Delta E = S \log \frac{C_x V_0 \pm C_s V_a}{(V_0 + V_a) C_x} \qquad (3.5)$$

For the KAM, $E_2 > E_1$ and for the KSM, $E_2 < E_1$, both for cations and anions. On taking antilogarithms, the equation

$$C_x = \pm \frac{C_s V_a}{10^{\Delta E/S}(V_0 + V_a) - V_0} = \pm \frac{V_a}{V_0 + V_a} \frac{C_s}{\left(10^{\Delta E/S} - \frac{V_0}{V_0 + V_a}\right)} \qquad (3.6)$$

is obtained. Assuming that $V_a \ll V_0$, it then follows that

$$C_x = \pm C_s \frac{V_a}{V_0} \frac{1}{10^{\Delta E/S} - 1} = \pm \frac{C_\Delta}{10^{\Delta E/S} - 1} \qquad (3.7)$$

where C_Δ is the change in the concentration due to the addition. Again, the $+$ sign refers to the KAM and $-$ to the KSM. The V_a, V_0 and C_s values are known. If the slope is not Nernstian and/or the temperature is not known accurately, the S value is replaced by the experimentally obtained value, S_{ex}. Usually the slope of the E vs. log a dependence is measured, obtained with standards at a known temperature. A simpler method of determining the slope involves reading the potential before and after dilution of the solution [26]. After the potential reading the solution is diluted with the blank solution from volume V_1 to volume V_2. Then

$$S_{ex} = \frac{\Delta E_{dil}}{\log V_2/V_1} \qquad (3.8)$$

where ΔE_{dil} is the difference in the potentials before and after dilution.

The concentration to be determined, C_x, can basically be obtained in three ways.

(1) By calculation from Eqs. (3.6) and (3.7).

(2) By using the tables which were first published by the firm Orion [27–29]. In these tables values of the C_x/C_Δ ratio are given for various ΔE and S (or S_{ex}) values. The C_x value is obtained by multiplying the ratio by the fraction $(C_s V_a)/V_0$. When the values are suitably selected, e.g. $C_s = 1$ and $V_0 = 100 V_a$, then the results are obtained from the tabulated ratios by mere change in the position of the decimal point. These tables are given in the Appendix for the KAM and KSM. The relationship for C_Δ is identical for the KAM and KSM only when the

species which removes the substance to be determined during the KSM reacts with it in a 1 : 1 stoichiometric ratio. Otherwise

$$C_\Delta = \mu \frac{V_a}{V_0} C_s \qquad (3.9)$$

where μ is the appropriate stoichiometric coefficient. For example, when the precipitation reaction between S^{2-} and Ag^+ is monitored with the sulphide-sensitive electrode, $\mu = 1/2$; when the silver-sensitive electrode is used, $\mu = 2$.

(3) The C_x value can also be obtained by using nomograms instead of tables, as has been shown by Karlberg [30]. However, if they were to be used widely, their accuracy would have to be improved.

The KAM has several advantages. It is more rapid than calibration with two standards (according to the literature [31], it requires only half the time necessary for the calibration) and it is often more precise ($\pm 2 \%$) [32]. It is possible to determine two or more species simultaneously if suitable electrodes are available and the ions do not mutually interfere, e.g. F^- and Cl^-; then the standard solution added contains the two species and the determination is very rapid (15 samples per hour [31]).

Its main disadvantage is that it is somewhat more complicated, which, however, is the case for all the addition methods. This drawback is considerably suppressed by using microprocessor-operated instruments. Methods involving a single addition are the least advantageous, as electrode calibration is necessary in order to determine the slope, S. The KAM is especially suitable for concentrations (activities) at least 100 times higher than the detection limit, where the E vs. log a dependence is linear and generally Nernstian, although it has also been proposed for lower concentrations [32]. In order to eliminate the effect of the ionic strength, a modified KAM has been developed [33], in which a mixture of a standard solution and of the sample solution is added to the measured solution. The KSM is much less extensively used in practice.

3.2.2.2 THE ANALYTE-ADDITION
AND SUBTRACTION METHODS

When the sample solution is added to a standard solution and the electrode potential is measured before and after the addition (E_1 and E_2, respectively), then

$$E_1 = E^{0'} \pm \frac{RT}{nF} \ln \frac{\gamma_1}{\alpha_1} C_s \qquad (3.10)$$

$$E_2 = E^{0\prime} \pm \frac{RT}{nF} \ln \frac{\gamma_2}{\alpha_2} \left(\frac{C_s V_0 \pm C_x V_a}{V_0 + V_a} \right) \qquad (3.11)$$

where the $+$ sign refers to analyte addition and the $-$ sign to analyte subtraction. These equations are analogous to Eqs. (3.2) and (3.3) with C_x and C_s interchanged. The ΔE value is then given by

$$\Delta E = S \log \left[\frac{C_s V_0 \pm C_x V_a}{(V_0 + V_a) C_s} \right] \qquad (3.12)$$

ΔE is positive for analyte addition and negative for analyte subtraction. Equation (3.12) can be rewritten in the form

$$C_x = \pm \frac{C_s}{V_a} [10^{\Delta E/S}(V_0 + V_a) - V_0] \qquad (3.13)$$

and for $V_a \ll V_0$

$$C_x = \pm \frac{C_s V_0}{V_a} [10^{\Delta E/S} - 1] \qquad (3.14)$$

When tables are to be used, the relationship $C_\Delta = C_x(V_a/V_0)$, is introduced, yielding

$$\frac{C_\Delta}{C_s} = \pm [10^{\Delta E/S} - 1] \qquad (3.15)$$

The evaluation is quite analogous to that for the KAM and KSM. For analyte subtraction, Eq. (3.9) holds. (For the tables see the Appendix.)

The methods of analyte addition make possible analyses of small sample volumes, without the use of microelectrodes, even if the macro-electrode would be insufficiently wetted. Moreover, any pH adjustments and masking of interfering species before the measurement of E_1 are unnecessary, as the standard can already be suitably prepared from this point of view. It is advisable to employ these methods whenever there is danger of chemical change in the solution on dilution. The analyte-subtraction method widens the scope of the measurement considerably, as substances not sensed by the electrode can be determined if they undergo a stoichiometric reaction with the species sensed by the electrode. A good example of this principle is the determination of residual chlorine (Cl_2, HOCl and OCl^-) and of organic chloro-derivatives in potable waters by monitoring the decrease in the activity of iodide ion, with the iodide-sensitive electrode [34].

3.2.2.3 DOUBLE-ADDITION METHODS

By two additions of a standard solution, three potential values can be obtained and hence three unknown variables are permissible, i.e. C_x, E° and S. The equations for E_1 and E_2 are identical with (3.2) and (3.3), respectively, and for equal additions potential E_3 is given by

$$E_3 = E^{0\prime} \pm \frac{RT}{nF} \ln \frac{\gamma_3}{\alpha_3} \left(\frac{C_x V_0 \pm 2C_s V_a}{V_0 + 2V_a} \right) \qquad (3.16)$$

The same assumptions hold here as in the single-addition methods. If ΔE_2 is given by Eq. (3.4) and $\Delta E_3 = E_3 - E_1$ (not $E_3 - E_2$) corresponds to the relationship

$$\Delta E_3 = E_3 - E_1 = S . \log \frac{C_x V_0 \pm 2C_s V_a}{C_x(V_0 + 2V_a)} \qquad (3.17)$$

then the ratio

$$R = \frac{\Delta E_3}{\Delta E_2} = \frac{\log \dfrac{C_x V_0 \pm 2C_s V_a}{C_x(V_0 + 2Va)}}{\log \dfrac{C_x V_0 \pm C_s V_a}{C_x(V_0 + V_a)}} ; \qquad C_\Delta = \frac{C_s V_a}{V_0} \quad \text{(for } V_a \ll V_0 \text{)} \qquad (3.18)$$

can be defined. The $+$ sign again holds for addition and the $-$ sign for subtraction. The equations for analyte double addition or subtraction differ from the equations above in that C_x and C_s are interchanged. However, double addition of the sample solution has little practical importance.

As the slope does not appear in Eq. (3.18), the R value is identical for ions of different charge. C_x can again be calculated or found in tables listing R for various C_x/C_Δ ratios, assuming that $V_a \ll V_0$. For the tables see the Appendix. The manual calculations are rather tedious; therefore an iteration method using a minicomputer has been developed [35]. In this computation, volume V_a is not neglected with respect to V_0. The method is not recommended for $C_x V_0 < C_s V_a$. Slow convergence is sometimes encountered even for $C_x V_0 > C_s V_a$.

Another method is based not on the addition of two identical volumes of a standard solution, but on addition of a standard solution followed by dilution in a known ratio. It is advantageous to dilute so that the original concentration of the species to be determined is restored. The change in the electrode potential due to dilution determines the electrode response slope according to Eq. (3.8) and the potential difference before and after the addition then determines C_x. When the dilution ratio is constant (e.g.

1:1), a table giving R values for various C_x/C_A ratios can again be drawn up [26], (see Appendix). If a buffer changing the amount of ions present (e.g. TISAB) is added before the measurement, the dilution must be carried out with a mixture of water and the buffer in the same ratio. This method is sometimes called the known-addition known-dilution method.

The main advantage of the double-addition methods is that they are the simplest procedures for obtaining C_{tot} and their main disadvantage is poor precision. Hence they should be applied chiefly in measurements close to the detection limit of the electrode.

3.2.2.4 MULTIPLE-ADDITION AND SUBTRACTION METHODS

Generally, for the $(m - 1)$th addition of the same volume of a standard solution,

$$E_m = E^{0\prime} \pm \frac{RT}{nF} \ln \frac{\gamma_m}{\alpha_m} \left[\frac{C_x V_0 \pm (m - 1) C_s V_a}{V_0 + (m - 1) V_a} \right] \qquad (3.19)$$

where the $+$ sign before the log term again holds for addition and the $-$ sign for subtraction. These methods yield the most precise results, but manual calculation of the concentration to be determined from a series of potential values read after the individual additions would be far too tedious. Therefore the evaluation is carried out either graphically or numerically by computer (see Section 3.2.4).

3.2.3 Potentiometric titrations

In many applications, direct potentiometry with an ISE is not sufficiently accurate and precise. Substantial improvement can be achieved, as in other potentiometric methods, by performing a titrimetric determination. Series of samples of similar concentrations are also better analysed by titration. For titrations with ISE's similar considerations are valid as for potentiometric titrations with indicator electrodes of the first kind and, in addition, the selectivity of the electrode must be taken into account. The titration is actually a form of the multiple-subtraction method (see Section 3.2.2.4).

In principle, any type of titration can be performed with an ISE. Acid–base titrations are the domain of pH-selective glass electrodes; the other ISE's are virtually exclusively used in precipitation and complexometric titrations, as suitable electrodes for systems which are best analysed by redox titration are not as yet available. Many titration procedures have already been described; a large proportion of them involve fluoride ions, owing to the outstanding properties of the lanthanum trifluoride single-crystal electrode. Most other titrations make use of copper-, sulphide- and halide-sensitive electrodes.

For the end-point of a titration to be determinable with an ISE, either the titrand or the titrant must be sensed by the electrode, for example, titrations of fluoride with solutions of Th(IV), La(III) or Ca(II) by using the fluoride-sensitive electrode, first performed by Lingane [36], titration of sulphate with Pb(II) and indication by the lead-sensitive electrode [37], indirect titration of Al(III) by using a copper-sensitive electrode [38], or else some kind of electrometric indicator, sensed by the electrode and taking part in the chemical mechanism of the titration, must be present (for example, ferric ion can be titrated with EDTA, by use of the fluoride-sensitive electrode and addition of a small amount of fluoride to the solution [39]). Sometimes, a single ISE can be used for monitoring the activities of various ions, e.g. the iodide-sensitive electrode will respond to mercuric, gold and cyanide ions or the chloride electrode to bromide and iodide and thus they can be employed in a great variety of titrations. Many chelometric titrations, for which ISE's cannot be used directly for indication, can be performed by employing the modified principle of electrometric indicators originally developed by Siggia *et al.* [40] and by Reilley and Schmidt [41] for chelometric titrations with mercury electrodes, based on the monitoring of the mercury electrode potential during the titration of the determinand metal in the presence of a small amount of the mercury(II) chelate with the titrant (see below). The latter authors have also described a simple method for prediction of the shape of titration curves from the pH vs. E relationship (the Pourbaix-Reilley diagram) [41].

To obtain satisfactory results, the titration conditions must be optimized, in order to satisfy both the requirements of the particular chemical reaction and those of the indicator electrode. The chemical reaction underlying the determination should have a well-defined mechanism, should not be complicated by side-reactions (or at any rate the effect of side-reactions, such as competing reactions of protons with the ligand in chelometric titrations, should be quantitatively describable and reproducible), and the equilibrium constant of the main reaction should be as large as possible in order to minimize the titration error. Simultaneously, the indicator electrode should respond specifically to a single species involved in the titration reaction, with sufficient accuracy, precision and sensitivity, should be indifferent to any other component of the system and should not be attacked by the titration system. Moreover, the chemical reactions involved and the electrode response should be sufficiently rapid. In titrations with an electrometric indicator, the conditions employed must further ensure that the substance functioning as the indicator does not interfere with the main reaction, but that it reacts with the titrant immediately after the reaction of the titrand is completed (i.e. the electrometric indicator forms a less stable complex or a more soluble precipitate with the titrant than that formed by the titrand, but the equilibrium constant of its reaction must still be high enough for the titration error to be as small as possible). On the other hand, when a metal complex with the titrant is used as the electrometric indicator (see Reilley and Schmidt [41]), it must be more stable than the complexes formed during the titration. However, it should be borne in mind that titrations with electrometric indicators are necessarily subject to a larger error than titrations in which the titrand or the titrant are directly sensed by the indicator electrode.

In an ideal titration, the end-point detected (located generally as the inflexion point of the titration curve) should be identical with the true equivalence point. However, in practice the end-point is more or less distant from the equivalence point. This is generally caused by several factors: (a) by the fact that the equilibrium constant of the reaction is not infinitely large, (b) by the effect of dilution of the titrand solution with the titrant, (c) by interferences with the electrode response by other species, (d) by errors in the location of the end-point.

Meites and co-workers [42−44] derived basic relationships for the location of the inflexion point in acid–base, precipitation and chelometric titrations, assuming an ideal indicator electrode, and showed that the effect of dilution generally causes the inflexion point to be different from the equivalence point. The error thus caused is, however, relatively small. The theoretical treatment of precipitation and complexometric titrations with ISE's [45−48] then indicated that interference with the electrode response and a low value of the equilibrium constant led to substantially larger titration error. For a titration based on a combination (complexometric or precipitation) reaction

$$M^{n+} + X^{n-} \rightleftharpoons MX \tag{3.20}$$

where MX is either a precipitate with the solubility product

$$K_s = [M^{n+}][X^{n-}] \tag{3.21}$$

or a complex with the dissociation constant

$$K_d = \frac{[M^{n+}][X^{n-}]}{[MX]}$$ (3.22)

then when, say, M^{n+} is monitored during the titration, Meites's treatment [43, 44] gives the following relationships for the concentration of M^{n+} at any point of the titration [45, 46]:

$$[M^{n+}] = \frac{\varphi + \sqrt{\varphi^2 + 4K_s}}{2} \qquad \text{(precipitation titration)}$$ (3.23)

$$[M^{n+}] = \frac{\chi + \sqrt{\chi^2 + \dfrac{4K_d C_M^0}{1 + rx}}}{2} \qquad \text{(complexometric titration)}$$ (3.24)

where C_M^0 is the initial sample concentration, x is the degree of titration, defined by

$$x = \frac{C_x V_x}{C_M^0 V_M^0}$$ (3.25)

where V_M^0 is the initial sample volume, C_x is the titrant concentration and V_x is the titrant volume required to reach the equivalence point, r is the dilution factor

$$r = \frac{C_M^0}{C_x}$$ (3.26)

and φ and χ are given by

$$\varphi = \frac{(1 - x)C_M^0}{1 + rx}$$ (3.27)

and

$$\chi = \frac{(1 - x)C_M^0}{1 + rx} - K_d$$ (3.28)

If the indicator electrodes used were perfectly specific, relationships (3.23) and (3.24) could be substituted directly into the logarithmic term of the Nernst equation and the titration curve calculated. However, actual electrodes suffer from interferences and their potential must be expressed by the relationship

$$E = E^{0\prime} + \frac{RT}{nF} \ln \left[a_M + \sum_i k_i^{Pot}(a_i)^{n/z_i} \right]$$ (3.29)

where a_i, k_i^{Pot} and z_i are the activities, selectivity coefficients and charges of the interfering ions, respectively. The interfering ions may be present in the sample, the titrant, or both. For the general case, when interfering ions are present both in the sample and in the titrant, the electrode potential is given by the equation [45, 46]

$$E = E^{0\prime} + \frac{RT}{nF} \ln \left\{ [M^{n+}] + \sum_i \frac{k_i^{Pot}C_i^0}{(1 + rx)^{n/z_i}} + \sum_i k_i^{Pot}C_i^{0\prime} \left(\frac{rx}{1 + rx} \right)^{n/z_i} \right\}$$

(3.30)

(assuming that the activity coefficients and liquid-junction potentials are constant and hence the concentration can be used instead of the activity), where C_i^0 and $C_i^{0\prime}$ are the initial concentrations of the interfering ions in the sample and in the titrant, respectively.

In order to find the inflexion point, the second derivative of Eq. (3.30) must be calculated and set equal to zero. Schultz [45, 46] and Carr [47, 48] carried out these computations in different ways and calculated theoretical titration errors for various sample concentrations, equilibrium constants, dilution factors and interfering ion concentrations. It was generally concluded that the titration error increases with decreasing sample concentration, increasing solubility product or dissociation constant, increasing dilution factor and increasing concentration of interfering ions (which is scarcely unexpected!). The errors increase when the titration reaction is non-symmetrical. Generally, the overall error is given by a combination of these factors: the greatest effect is exerted by the sample concentration and lesser effects by the equilibrium constant and the interfering ions; dilution has the least effect. In order to obtain errors below 1 %, it is generally necessary that $C_M^0 \geqq 10^{-2}M$, $K \leqq 10^{-8}$, $\sum_i k_i^{Pot}C_i^0 \leqq 10^{-3}-10^{-4}$ and $r \leqq 0.3$. The titration error can often be decreased by employing non-aqueous or mixed media, in which many precipitates are less soluble or complexes more stable, or where the electrode responds to lower concentrations (e.g. the solubility of the LaF_3 electrode membrane is suppressed). However, a disadvantage is that the electrode response is usually more sluggish in non-aqueous media (see e.g. [49]).

The actual titration error strongly depends on the method used for location of the end-point. This is mostly done graphically. As these procedures also have a certain application in the addition methods of direct potentiometry, they are treated separately in Section 3.2.4.

The principle of indirect indication in chelometric titrations of metals which are not sensed by the indicator electrode, based on the work of

Reilley and Schmidt [41] (see above), can be generalized as follows.
Let metal ion M be titrated with chelating agent Y (the charges on the ions are omitted and concentrations are used instead of activities for the sake of simplicity):

$$M + Y \rightleftharpoons MY; \qquad K_{MY} = \frac{[MY]}{[M][Y]} \qquad (3.31)$$

The indicator electrode is specific for another metal ion M':

$$E = E_{M'}^{0\prime} + \frac{RT}{nF} \ln [M'] \qquad (3.32)$$

Ion M' also forms a chelate with ligand Y, which is more stable than MY,

$$M' + Y \rightleftharpoons M'Y; \qquad K_{M'Y} = \frac{[M'Y]}{[M'][Y]}; \qquad K_{M'Y} > K_{MY} \quad (3.33)$$

Then, if a small amount of chelate M'Y is added to the sample solution, the electrode potential during the titration will be given by

$$E = E_{M'}^{0\prime} + \frac{RT}{nF} \ln \frac{[M][M'Y]K_{MY}}{[MY]K_{M'Y}} \qquad (3.34)$$

and hence the electrode will respond to variations in [M]. A very small concentration of M'Y (10^{-4}–$10^{-5}M$) is sufficient and need not be measured accurately. A copper (II)-sensitive electrode is particularly well suited for EDTA titrations of various metals, as Cu(II) forms a very stable EDTA complex (see e.g. [50, 51]).

3.2.4 Graphical and numerical evaluation methods

The multiple-addition methods (see Section 3.2.2.4) and potentiometric titrations require a procedure for finding the sample concentration from a series of electrode potential readings after multiple addition of a standard solution or a titrant. This can be achieved either graphically or numerically.

Reading of the end-point on a classical sigmoid titration curve is subject to a large error, especially when the curve is not symmetrical. Various graphical methods have been devised to improve the accuracy and precision. These methods were recently investigated [52]. It has been shown that the poorest accuracy and precision are attained when the end-point is located directly on the sigmoid curve, even if a graphical method, such as that of Tubbs [53] or Kohn and Zítko [54], is employed.

Better results are obtained by using derivative curves; when $\Delta E/\Delta V$ is plotted vs. V, the most important points lie around the equivalence point and these points are naturally subject to large errors, owing to the very low concentrations of the electroactive species in this region. Therefore Gran [55] proposed plotting $\Delta V/\Delta E$ vs. V, thus obtaining two approximately straight lines, the intercept of which is located on the volume axis and denotes the end-point. Here the points around the equivalence point may be neglected and the accuracy and precision improved; further improvement can be achieved by using the linear regression method for construction of the straight lines [52].

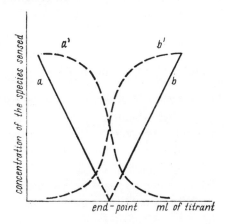

Fig. 3.1 Schematic representation of linearized titration curves. Substance A titrated with substance B. Curve (a) — substance A sensed by the electrode; curve (b) — substance B sensed by the electrode; curves (a'), (b') — the corresponding classical potentiometric titration curves.

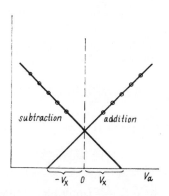

Fig. 3.2 The use of the Gran plot in the multiple-addition and subtraction methods.

In application of ISE's, the second Gran method [56] has found rather extensive use. Here the concentration of the species sensed by the electrode is plotted against the volume of the titrant or the standard solution. Therefore, concentrations must be calculated from the measured electrode potentials, corrected for dilution and plotted against the added solution volume. A straight line is obtained, the intercept of which with the volume axis denotes the end-point of the titration, or the unknown concentration in an addition method (see Figs. 3.1 and 3.2). The main disadvantage is the tediousness of calculating the volume corrections. The use of "Gran plot" paper [57], marketed by Orion Research Inc., makes the procedure simpler. Here, the Nernst equation

$$E = E^{\circ\prime} + S \log C \tag{3.35}$$

is linearized

$$\text{antilog}\,(E/S) = \text{antilog}\,(E^{\circ}/S) \cdot C \tag{3.36}$$

and E/S is plotted on a semi-antilog paper against the volume of solution added. The Orion paper has a built-in slope of 58 mV/pX, but can also be used for other integral slopes, e.g. 29 mV/pX. Further, the paper provides for correction of 100 % or 10 % dilution by skewing the co-ordinates upwards when proceeding from left to right.

The multiple-addition method can be evaluated analogously to titrations, as follows. When Eq. (3.19) is modified to

$$[V_0 + (m-1)\,V_a]\,10^{E_m/S} = \pm \frac{\gamma_m}{\alpha_m}\,10^{E^{\circ\prime}/S}[V_0 C_x \pm V_a(m-1)\,C_s] \tag{3.37}$$

the equation of a straight line, $y = aV + b$, is obtained, where

$$y = [V_0 + (m-1)\,V_a]\,10^{E_m/S} \tag{3.38}$$

$$a = \pm \frac{\gamma_m}{\alpha_m}\,10^{E^{\circ\prime}/S} \cdot C_s(m-1) \tag{3.39}$$

$$b = \pm \frac{\gamma_m}{\alpha_m}\,10^{E^{\circ\prime}/S} \cdot C_x V_0 \tag{3.40}$$

At the intercept with the x-axis (i.e. for $y = 0$)

$$\pm C_s V_x = -C_x V_0 \tag{3.41}$$

$$C_x = -C_s \frac{V_x}{V_0} \quad \text{(multiple addition)} \tag{3.42}$$

or

$$C_x = C_s \frac{V_x}{V_0} \text{ (multiple subtraction)}$$

where V_x is the volume of the standard solution added, corresponding to the concentration to be determined, C_x (see Fig. 3.2). The V_x value is negative for the addition method and positive for the subtraction method. For this method a computer-controlled apparatus was constructed and applied to the microdetermination of ammonia [57a].

Gran plots can also be employed in successive and indirect titrations. The disadvantages of Gran plots are that any departure of the electrode response slope from the theoretical value causes an error and must be corrected for, and the method does not provide for side-reactions. The method as modified by Ingman and Still [58] makes some provision for side-reactions, but has a disadvantage in that it requires accurate knowledge of the equilibrium constants and concentrations for the particular components taking part in the side-reactions. The Gran linearization method is suitable for determinations in the vicinity of the detection limit. Then the initial points depart from the straight line and the extrapolation leads to the sum, $C_x + C_l$, where C_l is the background concentration of the determinand, arising from impurities in the base electrolyte, dissolution of the membrane, etc. If the C_l value is known (e.g. a value of $6.3 \times 10^{-7} M$ is given for the solubility of the F^- membrane [59]), then the precision of the determination of C_x is better and C_x is given by

$$C_x = \pm(C_s + C_l)\frac{V_x}{V_0} - C_l \tag{3.43}$$

(The Gran titration can also be employed for the determination of the C_l value.)

The best precision and accuracy in evaluation of these linear plots can be achieved by numerical multiparametric curve fitting. These calculations are very tedious and require the use of a computer (see e.g. [35, 52]). Errors in the multiple-addition method were studied in detail by Buffle *et al.* [59−61] and the mathematical statistical method was compared with the graphical. In the statistical evaluation, possible non-linearity of the $y = f(V)$ relation caused by interactions of the medium with the species monitored or by sluggish response of the electrode need not be detected. In the graphical method, the initial points can be neglected and the V_x value determined from the points corresponding to higher V_a. However, linearity of the $y = f(V)$ function does not necessarily mean that a systematic error, due to variations in the ionic strength or to uncertainty in the determination of the electrode response slope, is absent. The best precision and accuracy are achieved when (*a*) the concentration of the standard solution in the addition method is as close as possible to that of the sample, (*b*) the electrode slope is known with good precision (better than 2 %), (*c*) $1 < [V_{a\,max}(C_s + C_l)]/[V_0(C_x + C_l)] < 3$, where C_l is the residual concentration given by impurities, membrane dissolution, etc. The precision increases with the number of additions; for very precise determinations the number of additions should exceed ten: hence the high precision of titrations.

3.2.5 Modified techniques

In order to increase the sensitivity and improve the precision and accuracy, various modified measuring techniques have been devised. The simplest way of lowering the absolute detection limit is to work on the microscale. Various modified electrodes have been developed for small volumes (see Chapter 2). Otherwise, the measuring technique remains the same.

Substantial improvement in accuracy and precision can often be achieved by using differential measuring techniques with two ISE's. Basically, two approaches can be taken [62].

(1) A cell without liquid-junction is employed, containing two different ISE's, one sensitive to the species to be determined and the other to another species, the activity of which is maintained constant. Then the latter electrode maintains a constant potential and functions as a reference electrode (e.g. the glass pH-electrode in an effectively acid–base buffered solution or the fluoride-sensitive electrode in solutions with constant F^- activity [63]).

(2) A cell with a liquid junction is employed, containing two identical ISE's, sensitive to the species to be determined, one immersed in the sample solution of concentration C_x and the other in a solution containing a known concentration, C_s, of the substance to be determined. Then the potential difference is measured, given by the relationship

$$\Delta E = \Delta E^{\circ\prime} + S_1 \log C_x - S_2 \log C_s \qquad (3.44)$$

(assuming that the liquid-junction potential is constant).

A technique employing the advantages of differential measurement and titrations, called null-point potentiometry [64−66], is especially useful. The experimental arrangement is identical with that described under (2); however, concentration C_s is not fixed but varied until $C_s = C_x$.

Then

$$\Delta E = \Delta E^{\circ\prime} + (S_1 - S_2) \log C \qquad (3.45)$$

With ideal electrodes ΔE should equal zero, since S_1 should equal S_2. However, in practice these values need not be identical and it is then necessary to determine the null-point voltage with the sample solution in both half–cells. This method is easily applicable to microanalyses (see e.g. [65]). It is convenient to generate the standard solution *in situ* coulometrically to eliminate corrections for dilution.

An advantage of differential techniques is that in approach (1) difficulties with variations in liquid-junction potentials and with blocking, leaking and maintenance of the reference electrode are avoided. On the other hand, the stability of the potential of electrodes of the second kind is usually better. Approach (2) substantially diminishes the effect of electrode-potential fluctuations. A disadvantage is that a circuit containing two ISE's may exhibit exceptionally large ohmic resistance, which places high demands on the measuring instrument and on the shielding. Brand and Rechnitz [62] designed a suitable differential amplifier for the purpose. A certain shortcoming of null-point potentiometry is that the null-point voltage need not be zero, owing to differences in the electrode response slopes. Wawro and Rechnitz [67] propose using split-crystal membranes to achieve better slope-matching: a crystal is halved and a hole is made through each half for passage of solutions. Each half then functions as a separate electrode of the differential pair. A differential potentiometric titration technique employing two ISE's with different selectivities (e.g. Ag_2S- and AgI-ISE's) has been developed [67a].

A technique similar to null-point potentiometry and to titration to the end-point potential was proposed by Weiss [68]. The potential of the ISE is measured in the sample solution; then the substance to be determined is chemically removed from the solution and a standard solution is added until the electrode potential reaches the original value. The method is simple in principle, but is more tedious than differential techniques and is feasible only with substances that can be simply and quantitatively removed from the solution.

3.3 CONTINUOUS AND AUTOMATED
MEASUREMENTS

Continuous and automated measurements are very important in practice, especially in production and pollution control. ISE's seem particularly well suited for these purposes; nevertheless many conditions must be met in order to obtain a reliable sensor, notably the following. (*a*) The electrode must respond selectively to the species required and must not be attacked by any other component of the system. (*b*) The electrode must measure reliably for prolonged time intervals, have a reasonably long life and be easily recalibrated when necessary. (*c*) The response time must be as short as possible. (*d*) The electrode must be robust. (*e*) The accessible concentration range must be sufficiently wide and the determination limit must be sufficiently low. Generally, these requirements are better met by solid-state electrodes, which are more robust, have longer lifetimes and especially shorter response time (often of the order of msec) than liquid ion-exchanger electrodes. Some liquid ion-exchanger electrodes, in particular the calcium-sensitive electrode, can also be used for continuous measurements, but it is necessary to choose the experimental conditions very carefully to achieve satisfactory results.

The simplest problem in this field is the application of ISE's for discontinuous measurements, advantageous for routine analysis of large numbers of similar samples. Here no special requirements are placed on the indicator electrode and the only difference is that manual work is excluded by using automated sampling, reagent delivery, and read-out. It is important that the programme should allow for sluggish response of the electrode and for it to recover between application of successive samples (see e.g. [69]).

In continuous measurements, all the conditions mentioned above must be met. The simplest way of measuring in flowing systems is to place the appropriate ISE and a reference electrode in the liquid stream. However, this system is rarely used in practice, as any change in the solution composition (pH, ionic strength, etc.) is likely to cause a change in the electrode potential. In practice a small portion of the sample is drawn from the stream and is mixed with the reagent(s) (e.g. TISAB) before it reaches the electrode system. Proportioning pumps are advantageously used for mixing the sample and reagents in the necessary ratios.

The use of conventional electrodes can be disadvantageous as discussed in Chapter 2 and Section 3.2.5. Therefore, differential measuring techniques are preferable. The same approaches as these described under (1) and (2) in Section 3.2.5 can be used. Two identical ISE's are placed at various positions in the flowing system (Fig. 3.3) [28]. Then it is possible to measure the following potential differences: $E_1 - E_2$, $E_1 - E_3$ and $E_3 - E_2$. When the reagent contains a known constant concentration of the species to be determined, the difference $E_1 - E_2$ yields the concentration directly, as electrode 1 acts as a reference electrode. The difference $E_1 - E_3$ gives the concentration by means of the analyte-addition method and $E_3 - E_2$ by the known-addition method. When the reagent contains a complexing agent, the difference $E_3 - E_2$ yields the sample concentration by the known-subtraction method. Substances for which no electrode exists can also be determined, if they can be complexed or precipitated by a substance sensed by the electrode and present in the reagent stream; then difference $E_3 - E_1$ yields the sample concentration by means of the analyte-subtraction method. Systems involving an electrode in position 3 have an additional advantage in that no calibration with a standard solution is required, provided that the electrode response slope remains constant. The flow in the tube not containing an electrode is stopped and any difference in the electrode potentials is cancelled by applying a suitable voltage (of course, the species to be determined must be present in the stream).

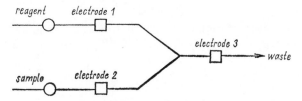

Fig. 3.3 Possible electrode positions in a simple flow-through system [28]. (By permission of Orion Research Inc.).

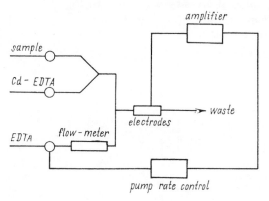

Fig. 3.4 Continuous monitoring of Ca^{2+} by means of titration with EDTA, using a Cd-sensitive electrode and Cd-EDTA electrometric indicator [28]. (By permission of Orion Research Inc.).

For the same reasons as those cited for discontinuous measurements (i.e. increased precision and accuracy), titration procedures are also introduced into continuous monitoring, though they have a serious disadvantage in the much more involved procedure employed (for an example of a continuous titration of calcium with EDTA, using a cadmium-sensitive electrode and the Cd-EDTA complex as electrometric indicator, see Fig. 3.4 [28]). An interesting approach to semi-continuous monitoring is gradient titration [70], which is analogous to classical titration except that the flow-rate of the titrant is constant and its concentration varies linearly; a suitable ISE is used to detect the end-point. The concentration gradient is produced by pumping the titrant at a constant rate into a vessel containing the solvent, where it is stirred and the solution pumped out. If C_T^0 and v_T^0 are the instantaneous concentration of the resulting solution and its pumping rate into the flow-through system, respectively, then

$$C_T = C_T^0 + (C_T^{init} - C_T^0) \left[\frac{(v_T^0 - v_T) t + V_T}{V_T^{init}} \right] \exp \frac{v_T^0}{(v_T - v_T^0)} \quad (3.46)$$

where C_T^{init} and v_T^{init} are the concentration and the volume of the resulting solution at time $t = 0$. For $v_T = 2v_T^0$, Eq. (3.46) simplifies to give

$$C_T = C_T^0 + (C_T^{init} - C_T^0)/V_T^{init}[V_T^{init} - v_T^0 . t] . e \quad (3.47)$$

and hence C_T is a linear function of time. Obviously it is possible to use varying concentration of the sample as well and to make gradients either increasing or decreasing with time. In Fig. 3.5 a scheme is given for the gradient titration of S^{2-} with Hg^{2+}, with use of a sulphide-sensitive electrode [70].

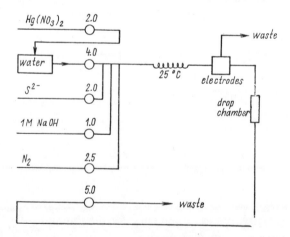

Fig. 3.5 Continuous gradient titration of sulphide with mercuric nitrate, using a sulphide-sensitive indicator electrode (by permission of the copyright holders, The American Chemical Society, from *Anal. Chem.*, 1974, **46**, 9).

A very useful and rapid method of automated analysis is the flow-injection technique, where the sample or a reagent is injected into the flow of a solution of constant composition. Reagent can also be generated coulometrically (see e.g. [70a, b]). For an automatic calibration of ISE's see e.g. [21a].

3.4 ION-SELECTIVE ELECTRODE MEASUREMENTS IN BIOCHEMISTRY, BIOLOGY AND MEDICINE

ISE's have found very wide application in biochemistry, biology and medicine, as is documented by hundreds of works scattered in the specialized literature. Specific features of this field will be briefly discussed here. A more detailed treatment can be found in several chapters of the book on glass electrodes edited by Eisenman [71 – 75], as most problems discussed there in connection with glass electrodes are generally valid for all ISE's. More recently, clinical applications of ISE's have been reviewed e.g. in [76–77b].

Plant and animal materials can generally be chemically pretreated (e.g. mineralized) and then the measuring technique does not differ from that of determinations in inorganic materials. However, this approach is tedious, time-consuming, brings the danger of contamination of the test material and, moreover, the biologist or physician is mostly interested in the concentration (activity) of a certain ion and its variations under conditions operative in a living organism. Therefore, the most common tasks are the determination of various ions (or non-ionic substances) in extracellular biological fluids, e.g. blood, plasma, cerebrospinal fluid, urine, gastric juices, bile, either *in vitro* or *in vivo*, and in intracellular liquids.

All biological fluids are very complex electrolytes, containing considerable amounts of proteins and other organic substances. The main ions are sodium, potassium, chloride and phosphate; an important role is also played by calcium, magnesium and bicarbonate ions. The pH of these solutions usually varies between 6.9 and 7.4, except for strongly acidic gastric juice, and urine, the pH of which can decrease below 4.5 [71]. It is thus evident that, from the biological and medical point of view, the most important inorganic levels to be determined are the pH, pNa, pK, pCa, pCl, pCO_2 and also pF, in view of the extensive fluoridation of potable waters. All these components can now be determined with commercial ISE's, the potentiometric method usually being more accurate and precise than the methods used previously (e.g. flame photometry for Na^+ [78]). Further, the use of ISE's is advantageous, since many biological processes depend on ion activities and not the overall concentrations. In modern biology and medicine, ISE's are successfully combined with radioactive tracer methods in kinetic studies and with absorption and emission spectroscopy in studying the bonding of ions in biological systems [72].

Of course, measurements with ISE's in these complex systems involve a number of problems. The presence of organic substances may considerably affect the liquid-junction potential values; this effect must be suppressed as much as possible by using suitable reference electrode liquid bridges (mostly chloride) and composition of standards for calibration of the indicator electrodes (mostly mixtures of KCl and NaCl with a suitable organic pH-buffer). The suspension effect may also play an important role (see Section 2.3.5) and the indicator electrode can be passivated by adsorption of organic substances (especially proteins); the sensitivity then decreases and the response time increases. Electrode passivation can sometimes be removed by soaking the ISE in suitable solutions, e.g. a pepsin solution in $0.1M$ HCl [74]; some authors have also attempted to suppress passivation by a suitable hydrophilic film on the electrode (see e.g. [79]). Sometimes, the surface layer of solid membranes must be removed, e.g. by grinding and polishing. On the other hand, ISE's can cause undesirable changes in biological fluids (e.g. haemolysis) or be toxic. From the point of view of the passivation and of the effect of the electrode materials, glass electrodes present the smallest problems, while some liquid ion-exchanger electrodes can be very toxic. During measurements *in vivo*, the problem of sterilization of the electrode system is also encountered.

Frequently, only small volumes of biological fluids are available and microelectrodes must be used. Intracellular measurements require very fine microelectrode systems (see Section 2.1.7). Silver chloride reference electrodes with chloride bridges are mostly satisfactory. In an attempt to decrease the liquid-junction potential value, ISE's selective for an ion having constant activity during measurement are often used as reference electrodes.

In measurements *in vivo*, flow-through systems are often employed (e.g. continuous measurements in the blood system). Cylindrical flow-through electrodes at which no undesirable turbulences are formed and reference electrodes with the lowest possible bleeding are the most suitable. A small part is usually taken from the main fluid stream and the measurement is performed in the side-stream. If the flowing fluid is under pressure (e.g. blood), the whole system must be properly sealed (e.g. by ground-glass joints) [72]. Continuous measurements *in vivo* are relatively easy to automate [72].

In discontinuous measurements on collected samples the sampling method is important. For example, in the determination of K^+ in plasma, the potassium content depends on the rate of sampling, as osmotic conditions vary with varying sampling rate. The pretreatment of the sample may also affect the results; in full-blood analysis, heparin is added, which partially complexes the ions present. This must be borne in mind, e.g. when the results obtained in blood and in plasma are to be compared. Constant experimental conditions must be carefully maintained, especially in the determination of CO_2 and ionized calcium, where constant pH and maintenance of equilibrium composition of the O_2–CO_2 atmosphere above the sample are essential, especially when the samples

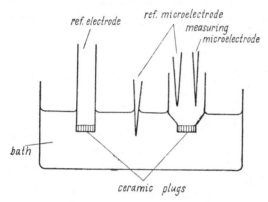

Fig. 3.6 Microelectrode calibration for intracellular measurements according to ref. [73]. For description see the text.

are stored [80]. An example of an automated system for analyses of samples of biological fluids is the "SPACE-STAT", developed by Orion Research for astronautical purposes [81].

Intracellular measurements with the microelectrodes described briefly in Section 2.1.7 are especially demanding. In principle it is possible to insert both an indicator and a reference electrode into a single cell and directly measure the particular ion activity, after calibrating the electrode with a standard with a composition as close as possible to that of the cell fluid. However, gross errors can be committed, as the reference microelectrodes have a high resistance, are easily blocked and exhibit a high liquid-junction potential (tip potential) of up to 30 mV [73, 74]. It is therefore safer to relate the indicator microelectrode potential to a reference macroelectrode immersed in a bath in which the test-cell is placed (Fig. 3.6). The following procedure is then employed for standardization [73]. The standard solution is placed in a cup, representing the test-cell, immersed in the bath and making contact with the bath solution through a porous porcelain plug in the bottom of the cup. The bath composition and the reference macroelectrode and microelectrode are identical during the standardization and the determination. Three potential measurements are carried out, with respect to the reference macroelectrode: 1 — for the reference microelectrode in the bath (E_1), 2 — for the reference microelectrode in the cup (E_2) and 3 — for the indicator microelectrode in the cup (E_3). The indicator microelectrode potential with respect to the reference macroelectrode is then $E_3 - (E_2 - E_1)$. The only changes in the liquid-junction potential occur at the tip of the reference microelectrode on immersion in the cup and on insertion into the cell; however, these changes are very small provided that the compositions of the standard and the cell fluid are similar.

REFERENCES

1. J. Doležal, P. Povondra and Z. Šulcek: *Decomposition Techniques in Inorganic Analysis*, Iliffe, London (1968).
2. J. C. van Loon: *Anal. Letters*, **1**, 6 (1968).
3. J. L. Guth and R. Wey: *Bull. Soc. Fr. Miner. Cristall.*, **92**, 105 (1969).
4. S. J. Haynes and A. H. Clark: *Econom. Geol.*, **67**, 378 (1972).
5. H. Chermette, C. Marfelet, D. Sandino, M. Benmalek and J. Tousset: *Anal. Chim. Acta*, **59**, 373 (1972).
6. N. A. Kazaryan and E. Pungor: *Acta Chim. Acad. Sci. Hung.*, **66**, 183 (1970).
7. P. J. Ke and P. J. Regier: *J. Fish. Res.*, **28**, 1055 (1971).
8. J. C. Warf, W. D. Cline and R. P. Tavebaugh: *Anal. Chem.* **26**, 342 (1954).
9. D. Weiss: *Chem. Listy*, **63**, 1152 (1969).
10. R. C. Rittner and T. S. Ma: *Microchim. Acta*, 404 (1972).
11. H. N. S. Schafer: *Anal. Chem.*, **35**, 53 (1963).
12. L. A. Elfers and E. D. Clifford: *Anal. Chem.*, **40**, 1658 (1968).
13. D. Weiss: *Chem. Listy*, **68**, 528 (1974).
14. E. W. Moore: *J. Clin. Invest.*, **49**, 318 (1970).
15. T. K. Li and J. T. Piechocki: *Clin. Chem.*, **17**, 411 (1971).
16. M. S. Frant and J. W. Ross, Jr.: *Anal. Chem.*, **40**, 1169 (1968).
17. J. Veselý: unpublished results.
18. D. Weiss: *Chem. Listy*, **65**, 305 (1971).
19. *Orion Res. Newsletter*, **III**, 35 (1971).
20. J. Veselý: *Collection Czech. Chem. Commun.*, **39**, 710 (1974).
21. F. Oehme and L. Doležalová: *Z. Anal. Chem.* **251**, 1 (1970).
21a. G. Horvai, K. Tóth and E. Pungor: *Anal. Chim. Acta*, **82**, 45 (1976).
22. R. A. Llenado and G. A. Rechnitz; *Anal. Chem.*, **44**, 1366 (1972).
23. A. Altinata and B. Pekin: *Anal. Letters*, **6**, 667 (1973).
24. T. B. Warner and D. J. Bressan: *Anal. Chim. Acta*, **63**, 165 (1973).

25. N. Shiraishi, Y. Murata, G. Nakagawa and K. Kodama: *Anal. Letters*, 6, 893 (1973).
26. *Orion Res. Newsletter*, II, 34 (1970).
27. *Orion Res. Newsletter*, II, 5 (1970).
28. *Orion Res. Newsletter*, II, 21 (1970).
29. *Orion Res. Newsletter*, I, 27 (1969).
30. B. Karlberg: *Anal. Chem.*, 43, 1911 (1971).
31. L. G. Bruton: *Anal. Chem.*, 43, 579 (1971).
32. H. J. Smith and S. I. Manahan: *Anal. Chem.*, 45, 836 (1973).
33. J. Buffle, N. Parthasarathy and D. Monnier: *Chimia*, 25, 224 (1971).
34. *Orion Res. Newsletter*, II, 26 (1970).
35. M. J. D. Brand and G. A. Rechnitz: *Anal. Chem.*, 42, 1172 (1970).
36. J. J. Lingane: *Anal. Chem.*, 39, 881 (1967).
37. J. W. Ross, Jr. and M. S. Frant: *Anal. Chem.*, 41, 967 (1969).
38. L. Šůcha and M. Suchánek: *Anal. Letters*, 3, 613 (1970).
39. H. Schäfer: *Z. Anal. Chem.*, 268, 349 (1974).
40. S. Siggia, D. W. Eichlin and R. C. Rheinhart: *Anal. Chem.*, 27, 1745 (1955).
41. C. N. Reilley and R. W. Schmid: *Anal. Chem.*, 30, 947 (1958).
42. L. Meites and J. A. Goldman: *Anal. Chim. Acta*, 29, 472 (1963).
43. L. Meites and J. A. Goldman: *Anal. Chim. Acta*, 30, 18 (1964).
44. L. Meites and T. Meites: *Anal. Chim. Acta*, 37, 1 (1967).
45. F. A. Schultz: *Anal. Chem.*, 43, 502 (1971).
46. F. A. Schultz: *Anal. Chem.*, 43, 1523 (1971).
47. P. W. Carr: *Anal. Chem.*, 43, 425 (1971).
48. P. W. Carr: *Anal. Chem.*, 44, 452 (1972).
49. E. Heckel and P. F. Marsh: *Anal. Chem.*, 44, 2347 (1972).
50. J. W. Ross, Jr. and M. S. Frant: *Anal. Chem.*, 41, 1900 (1969).
51. E. W. Baumann and R. M. Wallace: *Anal. Chem.*, 41, 2072 (1969).
52. T. Anfält and D. Jagner: *Anal. Chim. Acta*, 57, 165 (1971).
53. C. T. Tubbs: *Anal. Chem.*, 26, 1670 (1954).
54. R. Kohn and V. Zítko: *Chem. Zvesti*, 12, 261 (1958).
55. G. Gran: *Acta Chem. Scand.*, 4, 559 (1950).
56. G. Gran: *Analyst*, 77, 661 (1952).
57. *Orion Res. Newsletter*, II, 49 (1970).
57a. W. Selig, J. W. Frazer and A. M. Kray: *Mikrochim. Acta*, 675 (1975).
58. F. Ingman and E. Still: *Talanta*, 13, 1431 (1966).
59. J. Buffle, N. Parthasarathy and D. Monnier: *Anal. Chim. Acta*, 59, 427 (1972).
60. J. Buffle: *Anal. Chim. Acta*, 59, 439 (1972).
61. N. Parthasarathy, J. Buffle and D. Monnier: *Anal. Chim. Acta*, 59, 447 (1972).
62. M. J. D. Brand and G. A. Rechnitz: *Anal. Chem.*, 42, 616 (1970).
63. S. E. Manahan: *Anal. Chem.*, 42, 128 (1970).
64. H. V. Malmstadt and J. D. Winefordner: *Anal. Chim. Acta*, 20, 283 (1959).
65. R. A. Durst and J. K. Taylor: *Anal. Chem.*, 39, 1374 (1967).
66. R. A. Durst, E. L. May and J. K. Taylor: *Anal. Chem.*, 40, 977 (1968).
67. R. Wawro and G. A. Rechnitz: *Anal. Chem.*, 46, 806 (1974).
67a. I. C. Popescu, C. Liteanu and A. Clumocanu: *Rev. Roum. Chim.*, 26, 397 (1975).
68. D. Weiss: *Chem. Listy*, 66, 858 (1972).
69. I. Sekerka and J. F. Lechner: *Talanta*, 20, 1167 (1973).
70. B. Fleet and A. Y. W. Ho: *Anal. Chem.*, 46, 9 (1974).

70a. J. Růžička and E. H. Hansen: *Anal. Chim. Acta*, **78**, 145 (1975).

70b. G. Nagy, Z. Fehér, K. Tóth and E. Pungor: *Hung. Sci. Instr.*, **41**, 27 (1977).

71. E. W. Moore: in *Glass Electrodes for Hydrogen and other Cations. Principles and Practice*, (G. Eisenman, ed.) p. 412. Dekker, New York (1967).

72. S. M. Friedman: in ref. 71, p. 442.

73. J. A. M. Hinke: in ref. 71, p. 464.

74. R. N. Khuri: in ref. 71, p. 478.

75. P. Sekelj and R. B. Goldbloom: in ref. 71, p. 520.

76. N. Gochman and D. S. Young: *Anal. Chem.*, **45**, 11R (1973).

77. G. A. Rechnitz: *Am. Lab.*, **6**, 13 (1974).

77a. H. J. Berman, N. C. Hébert (eds.): Ion-Selective Microelectrodes, Plenum Press, New York (1974).

77b. J. Koryta, J. Pradáč, J. Pradáčová and M. Březina: Electrochemical Measurements in Vivo, in *Electroanalytical Chemistry*, A. J. Bard (ed.), Dekker, New York—in press.

78. E. W. Moore and D. W. Wilson: *J. Clin. Invest.*, **42**, 293 (1963).

79. M. A. Veksler, A. Y. Korolev and V. A. Golant: *Khim.-Farm. Zh.*, **2**, 56 (1968).

80. *Orion Res. Newsletter*, **III**, Nos. 9, 10 (1971).

81. *Orion Res. Newsletter*, **VI**, No. 2 (1974).

Chapter **4**

APPLICATIONS OF ION-SELECTIVE ELECTRODES

4.1 DETERMINATION OF FLUORINE COMPOUNDS

4.1.1 Determination of Fluorides

4.1.1.1 FLUORIDE-SELECTIVE ELECTRODE

The fluoride-selective electrode [1, 2] has found the widest application of all ISE's (except for the pH-electrode) because of its outstanding properties and because of the tediousness of other methods for the determination of fluorine compounds. The fluoride-ISE permits the determination of fluorine in practically any solid or liquid sample, provided that it is present as, or can be converted into, fluoride ion in solution.

Although it is well known that LaF_3, NdF_3, PrF_3, CeF_3 and CaF_2 can be employed as electroactive materials in fluoride-selective electrodes [1 – 3], only ISE's with an LaF_3 single-crystal activated by a small amount of EuF_2 are manufactured commercially. Relationships among the composition, crystallographic orientation and the function of the electrode have not yet been studied in sufficient detail [4, 5]; however, of these substances lanthanum fluoride probably gives the lowest detection limit and activation with Eu^{2+} is not indispensable. Polycrystalline membranes prepared by pressing and subsequent sintering (LaF_3 + 0.01–0.15 mole % EuF_2) have not yet been studied sufficiently [6] and probably cannot compete with single-crystal ISE's, e.g. in response rate and lifetime, but they are mechanically stronger [7]. Heterogeneous fluoride-sensitive ISE's are less sensitive [3]. Commercial ISE's have an internal reference solution with the $Ag/AgCl/Cl^- + F^-$ or $Ag/AgBr/Br^- + F^-$ reference electrode; solid-state electrodes have also been constructed [8]; they have substantially more negative potentials and exhibit poorer stability and reproducibility. Various modified electrodes can be used for microdeterminations. [9 – 11, 11a].

Lanthanum trifluoride crystallizes in a hexagonal system (the LaF_3 type, similar to Ce, Pr, Nd and Sm fluorides) [12] (see also [12a]). The rather complex lattice contains four non-equivalent F^- positions; the crystals are fragile and can break when exposed to sudden temperature changes of more than *ca.* 50 °C. The ionic conductivity of the crystal is rather high at laboratory temperatures ($10^{-7}\,\Omega^{-1}\,cm^{-1}$), which is surprising for a substance with a high melting point (*ca.* 1490 °C). The conductivity mechanism in LaF_3 is explained [2, 12a] in terms of a combination of neutral defects ([LaF_3] "holes"), which have a concentration of *ca.* 0.1 % at laboratory temperature [13], and LaF_3 units, according to the equation

$$[LaF_3] + LaF_3 \ \rightleftharpoons \ LaF_2^+ + F^- \qquad (4.1)$$

The particles formed, LaF_2^+ and F^-, each occupy the same volume as LaF_3. The charge is transported across the crystal by movement of F^- from LaF_3 to a neighbouring LaF_2^+ (the Schottky mechanism). On LaF_3-crystal activation (usually by less than 1 mole % EuF_2) the conductivity substantially increases [2, 4, 14]. Although there is no direct dependence between the resistance and function of the electrode, a considerable decrease in the response rate, reproducibility and, to a certain degree, in the sensitivity have been observed when the overall resistance of the membrane exceeds *ca.* 5 MΩ [4]. This value thus limits the choice of the shape and dimensions of the membrane.

4.1.1.2 PROPERTIES OF FLUORIDE-SELECTIVE ELECTRODES

It has been found that the electrode potentials may require several hours for stabilization at higher pH values, the shift being directed toward more positive values [15]. A more detailed study of the E vs. pH plot [5] indicates that prolonged potential variations can be observed in the whole pH range when the solution pH is changed with the electrode in the solution. These E vs. pH plots may exhibit different shapes for pH changes to higher or lower values, the effect lasting for several hours at pF > 4 (Fig. 4.1). These changes enclose a hysteresis region, dependent chiefly on the fluoride activity in the solution and the time elapsed since the pH change. For example, at a concentration of $10^{-6}M\,F^-$ [$I = 0.1(KNO_3)$, pH = 6.25 °C] the potentials can differ by 106 mV/min after a pH change and by 20 mV after one hour [5]. Hence, in analytical work the pH of the test solution must be adjusted in the absence of the electrode, if stable and reproducible results are to be obtained rapidly. However, in spite of the complicated behaviour during changes in the pH, the fluoride ISE is relatively stable under normal conditions (differences

of 0.5 mV in 8 hours and ± 3 mV over several months were observed
with Orion 94-09 ISE at 25 ± 0.1 °C [16]). The lowest fluoride activities can be determined at pH values within the range 5–6 (Fig. 4.1),
where the detection limit is *ca.* $10^{-7} M$ F$^-$ (0.002 ppm, 1.9 µg F/l.). It
is stated in the literature that this detection limit increases in solutions

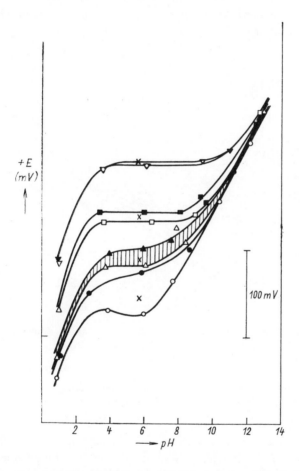

Fig. 4.1 The E vs. pH dependence for the 94-09A fluoride electrode (Orion), [5].
Silver chloride reference electrode, 25 °C; the ionic strength was adjusted with KNO$_3$.
The potentials were read 60 min after the pH-change or the electrode immersion. The
shaded area is the hysteresis region obtained in $10^{-6} m$ F$^-$. Solutions of pH ~ 1
and pH > 10.5 were prepared separately; otherwise, pH was adjusted by additions
of HNO$_3$ and KOH. ▽, ▼ — $10^{-4} m$ F$^-$; □, ■ — $10^{-5} m$ F$^-$; △, ▲ — $10^{-6} m$ F$^-$;
○, ● — $0.1m$ KNO$_3$; ▽□△○ — pH changed to lower values; ▼■▲● — pH changed to
higher values.
(by permission of Elsevier Publishing Co.).

containing anions of aminopolycarboxylic or carboxylic acids, especially citrate, by about one order of magnitude [17, 18], perhaps due to complexation of La^{3+} [17, 19]. Our experiments with various concentrations of citrate (Fig. 4.2), however, show that its presence only delays the electrode response at fluoride concentrations below 20 µg/l., the increase in the detection limit being relatively small ($2 \times 10^{-7} M$ F^- with potential reading after 20 min). An increase in the detection limit can also be caused by contamination with fluorides from the chemicals employed [16, 18, 20].

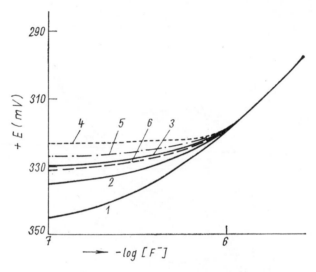

Fig. 4.2 The effect of sodium citrate and DCTA on the calibration curve of a Crytur 09-17 fluoride-ISE. (1) — $0.5M$ NaCl or $0.5M$ NaCl + $0.02M$ Na citrate, (2) — $0.5M$ NaCl + $5.7 \times 10^{-4} M$ DCTA, (3) — $0.5M$ NaCl + $0.2M$ Na citrate Curves 1, 2 and 3 were measured after 15 min. Curves 4, 5, 3 and 6 represent the dependence in $0.5M$ NaCl + $0.2M$ Na citrate, obtained after 5, 10, 15 and 20 min, respectively. The curves were obtained for stirred solutions at pH 6.1 and 22 °C.

The practical detection limit seems to be determined not only by the solubility of LaF_3 but also by accumulation of fluoride ions on the membrane surface (self-poisoning) [5, 22]. Recent results can better be explained by assuming a $pK_s(LaF_3)$ value of 29–30, corresponding to $8.3–4.2 \times 10^{-8} M$ activity of F^- in water, or even a lower pK_s value [22]; therefore it seems that there is actually a large difference between the solubility of freshly prepared LaF_3 ($pK_s = 17–19$) and crystalline LaF_3. This may be caused by the substantially lower surface energy of the crystal or its slow dissolution, causing the a_{F^-} value at the membrane to be determined by dissolution kinetics [23]. The inability of the electrode to respond to changes in $a_{La^{3+}}$ in the absence of a fine LaF_3 precipitate [5] and a considerable irreversibility in the formation and dissolution of colloidal particles containing La^{3+}, F^- and OH^- also indicate that the role of the solubility is not the same with LaF_3 membranes as with e.g. AgCl [22, 24, 25]. The solubility of LaF_3 is usually even lower in polar non-aqueous media (e.g. a lower detection limit was observed with polycrystalline LaF_3 in 80 % ethanol [6]), but the potential-stabilization in titrations in the presence of ethanol, methanol, 2-propanol and 1,4-dioxan is slower [26, 26a]. This is perhaps caused by slower diffusion in the stationary layer adhering to the electrode surface [27].

In the presence of ions binding fluoride in strong complexes, e.g. Th^{4+} or Zr^{4+}, a_{F^-} can be measured down to almost pF = 10 [21], but these measurements have no practical importance in direct potentiometry.

The electrode potential is, of course, affected by all species present and by reactions influencing the fluoride activity in solution. As poorly dissociated hydrogen fluoride and HF_2^- species are formed in acidic solutions, a_{F^-} decreases and the electrode potential shifts to more positive values, because the electrode responds to F^- even in strongly acidic media [30, 31]. However, no response to HF_2^- has been observed [32].

In the study of the equilibria

$$H^+ + F^- \rightleftharpoons HF(K_1) \tag{4.2}$$

$$HF + F^- \rightleftharpoons HF_2^-(K_2) \tag{4.3}$$

with a fluoride-ISE at 25 °C in various media, values of $\log K_1 = 2.89$–3.32 and $\log K_2 = 0.72$–0.86 were found [32, 33]. The best agreement with experimental values for pH > 1 and $[F^-] = 10^{-2} - 10^{-6} M$ was obtained for $\log K_1 = 3.02$, $\log K_2 = 0.73$ and $pK_s(LaF_3) = 28.54$ [34]. A maximum a_{F^-} value was found in the HF–H_2O system for 2 % w/w HF; with higher HF concentrations the fluoride activity decreases [35]. Here it should be pointed out that the glass electrode can be employed for pH-measurement in solutions with HF concentrations not exceeding $5 \times 10^{-3} M$, i.e. down to pH 2.1 in $10^{-2} M$ NaF, pH 4.1 in 0.1M NaF, or pH 4.9 in 1M NaF [36]. The pH of acidic fluoride solutions is measured with the quinhydrone electrode [36] and in industry sometimes with an antimony or other electrode [37].

Potassium fluoride is preferred for preparation of standard solutions, as it is more soluble than sodium fluoride (a saturated NaF solution is *ca.* 1M at 25 °C) and has a lower tendency to form ion-pairs, although NaF itself does not do so to a great extent [28]. In view of complex formation between La^{3+} and NO_3^- (the formation constant for $La(NO_3)_2^+$ is about 30 [29]) it is evidently better to adjust the ionic strength with potassium chloride, although nitrates have so far been used most frequently.

4.1.1.3 SELECTIVITY OF THE DETERMINATION

The most important interference is caused by the presence of OH^- ions, although it can be eliminated simply by adjusting the pH to below 6. The slope of the OH^--response increases more rapidly with increasing temperature than the Nernstian slope, but it does not attain the theoretical Nernstian value even at 60 °C and depends on the experimental conditions, owing to slow potential–stabilization in alkaline solutions. A slope of 33 mV/pOH was found from two potential values read after 60 min at pH 6 and 12 in $0.1M$ KNO_3; substantially higher values are obtained when the potential is read immediately after a pH change [5]. It can be considered proved that the mechanism of the OH^--interference is not simple blocking of the electrode surface by $La(OH)_3$, as exposure of the electrode to more concentrated hydroxide ($0.1M$ KOH) does not cause irreversible poisoning according to the equation

$$LaF_3 + 3\,OH^- \;\rightleftharpoons\; La(OH)_3 + 3\,F^- \qquad (4.4)$$

and the potential does not shift to negative values before stabilization, as is predicted by this equation. No chemical bonding of OH^- ions was indicated by spectroscopic and radiochemical methods [5, 39]. The electrode response to hydroxide, and some other electrode properties, can best be explained by assuming the existence of a surface film containing F^-, H_2O, OH^-, NO_3^-, FHF^-, HF and similar species bound predominantly electrostatically, possibly with participation of hydrogen and oxygen bonds [5, 24, 40].

In practical analyses it is mostly necessary to add masking buffers (see below) to the samples, usually containing anions of carboxylic acids, especially citrate and acetate, which, however, make the ISE response sluggish (Fig. 4.2). The effect of these anions (A) is not as pronounced in direct potentiometry [38, 39] as in titrations with La^{3+} ions, where precipitates of the $LaF_{3-x}A_x$ type ($0 < x < 1$) are formed [19]. After such titrations the electrode exhibits deviations from Nernstian function and sluggish potential-stabilization and has to be regenerated by polishing with an abrasive paste.

The most serious interference encountered in natural materials is from aluminium, as this is an element occurring very extensively and forming very robust complexes with fluoride. In contrast to common interferents, such as $Ca^{2+}, Mg^{2+}, Fe^{3+}$ and Ti^{4+}, Al^{3+} cannot usually be quantitatively masked with respect to F^- by addition of complexing agents, thus causing a negative error in the determination of fluoride that depends on the Al^{3+} concentration, the $Al^{3+} : F^-$ ratio, the kind and concentration of the complexing agent and the test-solution pH. The best results are obtained at complexing agent concentrations above $0.1M$ and pH values in the region of incipient Al^{3+} precipitation i.e. around pH 7, concentrations of fluoride of the order of $10^{-6}M$ then being measurable. Sodium citrate gave reasonable results in masking aluminium in concentrations up to 10 times that of the fluoride [15]; masking of 50 ppm of Al^{3+} in presence of 10 ppm of F^- by $0.5M$ sodium citrate at pH 6.0 was also reported [41]. In an effort to remove the adverse effect of citrate on the electrode response, the original composition of the TISAB buffer [42], which was found unsuitable for masking Al^{3+} [20], was changed by replacing citrate by 1,2-cyclohexanediaminetetra-acetic acid (DCTA) and the new mixture employed e.g. in the determination of fluode in waters [18, 43]. This buffer (B 1 in Table 4.1) is reported to mask 4 mg of Al^{3+} in the presence of ca. $10^{-6}M$ F^- without affecting the electrode response [20]. However, complexation between Al^{3+} and DCTA is rather slow and takes up to 20 hours [43]. In our experiments we did not succeed in quantitatively masking Al^{3+} concentrations higher than

Table 4.1

Buffers recommended for fluoride determinations in natural materials

No.	Composition	Resultant pH	Dilution ratio, buffer : sample	Ref.
B 1	$1M$ NaCl, $1M$ Na acetate, 0.014M DCTA	5.0—5.5	1 : 1	[18]
B 2	$1M$ NaCl, $1M$ Na citrate (neutral), 0.06M DCTA	6.0	2 : 1	[44]
B 3	1.25M NaCl (or NH_4NO_3), 1.36M Na citrate (neutral), 0.06M DCTA	6.5	1 : 10	—
B 4	$1M$ Tiron, 0.2M sulpho-salicylic acid, 0.5M EDTA (or DCTA)	6.0—6.5	1 : 10	[46]
B 5	0.1M Na_2HPO_4, 0.04M citric acid	6.3	1 : 1	[71]

0.5 mg/l. even after 20 hours (see Fig. 4.3). Substantially higher efficiency is obtained with a buffer containing a higher DCTA concentration and citrate (B 2 in Table 4.1) [44] and with buffer B 3 (see Fig. 4.3). The interference from Al^{3+} can also be suppressed by suitable dilution of the sample, but, of course, the fluoride concentration must not decrease below the electrode detection limit and a high concentration of the masking agent must be maintained [44, 45]. Very effective masking of Al^{3+}, as well as of Ti^{4+} and Fe^{3+}, can be attained with a buffer containing 10 % of 4,5-dihydroxybenzene-1,3-disulphonate (Tiron) at pH 7 [46−48], which permits the determination of *ca.* $10^{-6} M$ F^- in the presence of $10^{-2} M$ Al^{3+} (Fig. 4.3). The efficiency of this buffer is further improved by combining Tiron with sulphosalicylic acid and EDTA (buffer B 4, Table 4.1), so that Ca^{2+}, Mg^{2+} and UO_2^{2+} are also masked. However, the individual components of the buffer must be added to the test solution separately, first Tiron, then sulphosalicylic acid with simultaneous neutralization of the NaOH solution to pH 6.5 and finally EDTA. If the previously prepared buffer is added to the test solution, the masking effect appears as late as after 2 hours, apparently because of competition among EDTA, Tiron and fluoride for complexation of Al^{3+} [46]. We verified the masking effect of Tiron (see Fig. 4.3) and found it the best substance for masking Al^{3+} in the presence of fluoride; it would be advantageous to use it instead of citrate buffers, which are commonly employed.

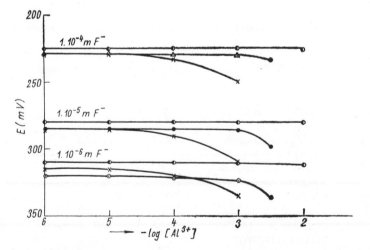

Fig. 4.3 The effect of buffers on Al^{3+} masking in the determination of fluoride. X — buffer B 1 (Table 4.1), ○ — buffer B 3 (Table 4.1), ◑ — 3 % Tiron, pH 7.0. Stirred solutions, 23 °C, the potential values read after 10 min, Crytur 09-17 electrode.

4.1.1.4 TITRATION DETERMINATION

According to Lingane [49], titration of fluoride with use of the F^--ISE for end-point detection is best carried out in an unbuffered medium with an $La(NO_3)_3$ standard solution. The titration curves are non-symmetrical and the potential-stabilization sluggish, especially around the end-point. Therefore, the author recommends titration to a preset potential in the presence of 70 % ethanol in the test solution [27]. Some authors recommend the traditional $Th(NO_3)_4$ as titrant, using a medium of 50 % ethanol or $0.01 M$ HNO_3 plus 80 % ethanol, which is reported to give better-developed breaks at the equivalence point, so that the fluoride detection limit decreases to 3.8 mg/l. compared with 19 mg/l. in the titration with the lanthanum salt [50 – 52]. However, the disadvantage of the titration with the thorium salt is slower potential-stabilization [53], the formation of colloidal ThF_4 [49] and film-formation on the indicator electrode [54], as well as on the liquid junction of the external reference electrode [53]. Fluoride can also be titrated with Al^{3+}, Fe^{3+}, UO_2^{2+}, Be^{2+}, Ca^{2+} and Zr^{4+}, but titration with the rare-earth ions seems to be most suitable [54]. The titration is carried out at pH 4–7 in a pyridine–hydrochloric (perchloric) acid buffer. The F^--ISE can also be employed in coulometric titration of fluoride (2–20 mg/l.) by anodically generated La^{3+} ions [55, 56]. Down to 0.5 µM F^- can be titrated with $0.001 M$ La^{3+} in polar solvents containing 5 % water [26]; the best results are obtained at pH 6.5 in the presence of glycine, which permits a wider choice of solvents. The presence of ammonium salts is undesirable because of the danger of hydrolysis and the formation of undissociated fluorides or compounds with La^{3+}. Because of slow electrode response during fluoride titration in polar solvents, a graphical interpolation method employing "time response graph paper" is proposed.

4.1.1.5 DETERMINATION OF FLUORIDE IN WATERS, INORGANIC SOLUTIONS, THE ATMOSPHERE AND GASEOUS MIXTURES

The fluoride-ISE is suitable for determinations in all kinds of water. At least 100 ml of water is sampled into polyethylene or polypropylene bottles and no preservative is added. In series of water analyses the calibration curve method is usually employed and fluoride concentrations of 20 µg–500 mg/l. are determined with a relative error from 0.3 to 10 %. A procedure for the assessment of concentrations below 2 µg/l. has been described [57]. In analyses of very pure waters or rain waters, in which concentrations of Fe^{3+}, Ca^{2+}, Mg^{2+}, and especially Al^{3+} above $10^{-6}M$ need not be expected, KNO_3 or preferably NaCl is added to a sample aliquot, so that the resultant salt concentration is at least $0.1M$. If necessary, the pH is adjusted to 5–6 with dilute HCl or KOH solution before the ISE is immersed. In this way concentrations close to the detection limit (2 µg/l.) can be determined. To natural surface, spring, industrial and potable waters with a low degree of mineralization and pollution (especially with Al^{3+} concentrations below 0.5 mg/l.), buffer B 1 is added (see Table 4.1). When a higher masking effect for Al^{3+} is required, buffer B 2 or B 3 is suitable; buffer B 3 is advantageous in geochemical prospecting for fluoride in surface waters [58]. The same buffers are suitable for the determination of fluoride in mineral, mine and deep waters, as well as in sea-waters.

In analyses of heavily polluted waters (e.g. waste waters) with complex and/or variable composition it is usually impossible to prepare satisfactory simulated standards and a standard addition method is used. Interfering cations can be extracted from such waters with 8-hydroxy-quinoline into a mixture of ethylene glycol monobutyl ether and chloroform; a glycine buffer with pH 10.5–10.8 is employed. The fluoride is then measured in buffer B 1 (Table 4.1) [59]. On the other hand, fluoride can be extracted into carbon tetrachloride containing triphenylanti-monyl(V) dichloride from a phosphate buffer of pH 4.0–6.5 containing DCTA. The fluoride is then back-extracted into dilute ammonia solution and the solution is neutralized with acetic acid. The method permits the determination of 10^{-1}–$10^{-8}M$ fluoride [60]. For heavily polluted waters preliminary separations are used involving distillation of HF [61] and ion-exchange [51]. Procedures involving preliminary separations are more tedious, but lead to an improvement in the precision of the determination (maximum relative error 1 %) and to a decrease in the detection limit because of simultaneous preconcentration of fluoride.

The determination of fluoride in inorganic salts, metal-plating baths, inorganic acids, etc., resembles that in strongly mineralized waters. Acidic or alkaline solutions require preliminary neutralization before addition of the buffer.

In determinations of fluoride in the atmosphere and gaseous emissions, the sample is passed through traps containing water [62] or an absorption solution, e.g. a mixture of K_2CO_3 + H_2O_2 [63], or directly through a buffer (e.g. B 1) [64]. Chemisorption cellulose filters impregnated with K_2CO_3, CaO or sodium formate [65–67], and membrane filters are also sometimes employed. For a more effective deposition of gaseous fluorine compounds from the atmosphere, sorption of F^- at an elevated temperature in a glass layer has been proposed [66]; the silicon tetra-fluoride formed is hydrolysed on contact with water and fluoride is formed. The determinable concentrations of fluoride in gaseous samples are of the order of $\mu g/m^3$. Flow-through continuous analysers have been proposed for automated monitoring of the fluoride concentration in gases [68, 68a].

4.1.1.6 DETERMINATION OF FLUORIDE IN MINERAL RAW MATERIALS

Inorganic materials which are incompletely soluble in water are dissolved in a mineral acid (e.g. phosphates), or alkaline [45, 69−73] or alkaline-oxidative [44] fusion, sintering [41, 73] and pyrohydrolysis [58, 74] are used. In prospecting for CaF_2 in rocks and soils the samples are extracted with a $Be(NO_3)_2$ solution acidified with HCl. The CaF_2 is dissolved by the conversion reaction; addition of buffer B 5 (Table 4.1) prevents formation of Be–F complexes and the liberated fluoride is measured directly [72]. For the extraction, $FeCl_3$ [72a] and $AlCl_3$ [72b] can also be used. Alkaline fusion and sintering with added Na_2CO_3 [41, 73, 75−77] are suitable for materials with higher contents of Ca, Mg, Fe and especially Al, as these elements remain quantitatively or almost quantitatively in the insoluble carbonate residue. The supernatant liquid is neutralized by hydrochloric acid and one of buffers B 1–B 5 (Table 4.1) is added. The solution must be neutralized at an elevated temperature and in the presence of e.g. citrate or DCTA in low concentration, to prevent formation of CO_2 bubbles on the ISE surface and precipitation of hydroxides, especially $Al(OH)_3$, on which fluoride is sorbed. After pyrohydrolytic sample decomposition or after distillation separation of HF, no buffer need be added to the solution, since this is a pure fluoride solution; a suitable indifferent electrolyte (NaCl, KNO_3 or NH_4Cl) is added and the pH is adjusted to 5–6 (acid–base indicator) [58].

4.1.2 Determination of Fluorine in Organic Substances

Similar techniques are employed as with inorganic materials; in view of interference of macromolecular compounds with the ISE function, addition techniques are more frequently used. If calibration curves are employed, simulated standards must be prepared and the influence

Table 4.2

Survey of application of fluoride-ISE's in the determination of fluoride
in inorganic substances

Material	Ref.	Material	Ref.
rain water	57, 78	in geochemical prospecting	72, 77, 107
potable water	51, 79—85	steel-mill slags	74, 109
surface waters, other		glass	44, 45, 71, 74
natural waters	18, 43, 59	enamels and paints	91
sea-water	86—89	welding and soldering fluxes	74, 110
waste waters	43, 59, 90	tungsten	111
H_3PO_4	91—93	selenium	112
HNO_3, other mineral acids	94, 95	aluminium (production	
fluorosilicic acid	96	control)	47, 113
HF production control	63	aluminium reductants	70
chromium-plating baths	97—99	oxides	44, 114
pickling baths for stainless		$CaSO_4$	63, 91
steel	100	coal	115
reduction cell electrolytes	70, 76	mineral fluorides	43, 69
nuclear fuel regeneration baths	101	cryolite and AlF_3	70, 76
metal-plating baths	102	NaF tablets	116
halogen phosphates	103, 104	ores	44
superphosphate production		minerals, rocks,	58, 69, 73, 77,
control	93	silicates	108, 117, 118
soils	72, 77, 105—107	phosphates and	41, 44, 45, 91,
		phosphate rocks	93, 119—123
clays	108	smoke and waste gases	62, 63, 67, 68,
			124—126
stream sediments	44, 72	atmosphere	57, 62, 64, 67,
			78, 80, 127, 128

of the organic matrix decreased by diluting the test solution and adding a suitable buffer [129] or NaCl to adjust to isotonic salinity in all solutions.

Fluorine bound in organic compounds must be converted into the ionic form, e.g. by digestion with perchloric acid on a water-bath as in the determination of fluorine in urine: the concentration of fluoride ions is measured before the digestion; after the digestion the solution is neutralized, a citrate buffer of pH 6 is added and the measurement is repeated. The concentration of bound fluorine is calculated from the difference of the two measured values [130].

In analysis of blood serum and plasma the sample is treated with heparin, centrifuged, mixed with an acetate buffer of pH 5.2–5.5, e.g. with buffer B 1 (Table 4.1), and measured [131]. For the determination of the overall fluorine concentration in blood, the sample is combusted at 600 °C in the presence of $MgCl_2$ and Na_2CO_3 as a fixative. Fluoride is separated from the ash by the diffusion method, the pH of the solution obtained is adjusted and the measurement is carried out [132].

Solid organic and biological materials insoluble in water are either decomposed by mineral acids (e.g. bones, teeth) or are combusted at 550 °C and the ash is dissolved in an acid [133, 134]. The rest of the analysis is then identical to that for mineral materials, e.g. phosphates. Solid organic samples can further be decomposed similarly to inorganic materials, e.g. by fusion with Na_2CO_3 [115], fusion with Na_2O_2 in a microautoclave [135], sintering with a mixture of $Na_2CO_3 + ZnO$ [136], pyrohydrolysis, etc. However, the more universal and elegant method of oxygen-flask combustion according to Schöniger [137, 138] is now more often used for mineralization of organic samples; the Kirsten method [139] of sample combustion in a tube with an oxygen atmosphere and the Wickbold method of combustion in a hydrogen-oxygen flame, which has been successfully applied especially to pharmaceuticals [140], are also used. In these combustion methods, the gaseous products are absorbed in NaOH or directly in a suitable buffer.

Fluorine can also be determined in organic fluoroderivatives by using a fluorine hydrogenation procedure [124]. The sample components are first separated by gas chromatography with hydrogen as carrier-gas, then led into a platinum tube heated to 1000 °C in which hydrogenation takes place, and the HF formed is absorbed in a suitable solution which is then introduced into the measuring cell. The procedure can be fully automated. Fluorine extraction from oil products is also based on hydrogenation [141]. The dried sample is mixed with sodium biphenyl in toluene under a nitrogen atmosphere. A small amount of water is added and fluorine is thus hydrogenated by nascent hydrogen. The hydrogen fluoride is then extracted with water from the reaction mixture, the aqueous phase is neutralized with nitric acid, a citrate buffer of pH 6 is added and the measurement is carried out. The method permits the determination of 0.01–25 ppm fluorine.

Many organic substances can be decomposed by treatment with mineral acids with simultaneous distillation of HF, by using the Willard and Winter method [61]. For example, in analysis of fish proteins [142] the sample is first digested with $HClO_4$ in the presence of $AgClO_4$ as a catalyst and HF is then distilled off at 140–150 °C. The distillate is neutralized with NaOH, buffer B 1 (Table 4.1) is added and the measurement is carried out.

Table 4.3

Survey of application of fluoride-ISE's in the determination of fluorine in organic substances

Material	Ref.	Material	Ref.
volatile fluoro derivatives	124, 152	wine	167, 168
fluoroborates	135, 140, 143	milk	169
trifluoromethyl compounds	144	human milk	170
organophosphates	135, 145	blood serum and plasma	131, 132, 154,
sulpho derivatives	146		171—175
organometallic compounds	147	saliva	176—178
biological materials	140, 142,	urine	130, 179—182
	153—157	faeces	183
pharmaceuticals	140, 158, 159	bones	133, 134, 184—186
vegetation	160 162	tissues	187
foodstuffs	163, 164	teeth and tooth enamel	133, 186—190
sugar-cane	136	toothpaste	191, 192
vitamins	159	mouthwash	193
fish and fish products	165	stains	194
beverages	166	household products	194
organic raw materials	148—151	oil and oil products	141

4.2 DETERMINATION OF CHLORINE COMPOUNDS

4.2.1 Determination of Chloride

4.2.1.1 CHLORIDE ELECTRODES

Potentiometric titration and direct determination of chloride are very old methods [195 – 198], but silver or silver chloride electrodes used for this purpose up to the middle of the nineteen-sixties are usually not included among ISE's. Direct determinations with silver chloride electrodes [196 – 198] are comparable with those employing ISE's based on AgCl [199], but the former may be subject to interferences from redox systems [200, 201]. For example, the presence of $10^{-3}M$ ferricyanide causes a shift in the silver chloride electrode potential of $+13$ mV in $0.1M$ KCl, while the potential of the single-crystal AgCl membrane remains unchanged [201]. All electrodes based on insoluble silver salts are affected by substances capable of reducing them to metallic silver, e.g. $SnCl_2$ or formaldehyde; the pH is frequently important. Membranes containing silver are less redox-sensitive than those containing mercury [201a]. Chloride-ISE's can be classified as:

(a) homogeneous electrodes containing an AgCl single-crystal [34, 201 – 203] or some other form of AgCl [200, 204; 204a];

(b) homogeneous electrodes with membranes containing a mixture of chloride and a chalcogenide of silver [205 – 207] or mercury [208, 208a, 209];

(c) heterogeneous electrodes with AgCl dispersed in silicone rubber [210 – 212] or in other plastics [213 – 215], or containing an inorganic ion-exchanger (hydrated ZrO_2) [216];

(d) LISE's with dimethyldistearylammonium chloride [17], tri-n-octylpropylammonium chloride [217] or the chloride salt of Aliquat 336S (methyltricaprylammonium chloride) [218] as electro ctive substances.

Chloride can also be determined indirectly with the Ag_2S electrode, by adding the sample to a standard silver solution (analyte addition) [219]. Chloride microelectrodes have also been constructed for test-solution volumes of more than 0.05 μl [220, 220a]; solid-membrane $AgCl/Ag_2S$ microelectrodes are more selective than LISE's [220b].

The detection limit of the AgCl-based ISE is chiefly determined by the membrane solubility, although interfering ions, adsorption and contamination of reagents with chlorides may also play a role [25]. From the solubility product ($K_s(AgCl) = 1.56 \times 10^{-10}$; $pK_s = 9.81$ at 25 °C [221]) it follows that, at equilibrium, $a_{Cl^-} = 1.25 \times 10^{-5}M$, but there seem to be small differences in the detection limit among individual ISE's. For $Ag_2S/AgCl$ membranes, a detection limit of $5 \times 10^{-5}M$ (1.80 mg/l.) [23] or 0.5 mg/l. [208a] has been given; a value of $10^{-5}M$ was reported for single-crystal membranes [203]. By tempering AgCl single-crystals at 320 °C for 40 hours in a partly evacuated space, the conductivity can be increased, differences in the conductivity of individual membranes substantially decreased and a somewhat lower detection limit attained [203]. Rapid cooling fixes an "excited state" in the single-crystal, which is perhaps maintained by exchange reactions taking place during measurements, as unused single-crystal electrodes age twice as rapidly [34].

The solubility of AgCl is rather high from the analytical point of view and therefore other substances have been sought. An HgS/Hg_2Cl_2

mixture, e.g. 60 mole % Hg_2Cl_2 + 40 mole % HgS [208], pressed at increased temperature exhibits Nernstian response within a chloride concentration range of 3500 mg–50 µg/l. and the detection limit is $5 \times 10^{-7} M$ Cl$^-$ (17 µg/l.) [208, 208a]. Silver chloride containing 10 mole% of colloidal gold also exhibits increased sensitivity [204, 204a]. In view of the high gold content in the latter membrane and the negligible resistance of the former [208], the question of their redox sensitivity arises. Later workers [221a] were unable to reproduce the Nernstian slope of the HgS/Hg$_2$Cl$_2$-ISE. Concentrations lower than $10^{-5} M$ Cl$^-$ can also be determined by adding certain polar non-aqueous solvents to the test solution or by measuring directly in non-aqueous media. The detection limit was found to be 10^{-7}–$10^{-8} M$Cl$^-$ in 1-propanol [222] ($pK_s = 14.36$ at 20 °C [223]), but the slope found with the Radelkis crystalline electrode in an organic phase is only 35–50 mV/pCl [213, 222]. A double-junction reference electrode is, of course, required in determinations of low chloride concentrations, to prevent contamination of the test solution with chloride.

The question of photosensitivity of silver halide membranes is not yet clear. With Ag$_2$S/AgCl membranes and with pressed AgCl in 0.1M NaCl, the maximum potential differences on change from darkness to normal laboratory light amounted to 1.5 and 30 mV, respectively [206]. However, a change of direct irradiation of a single-crystal membrane from 400 to 1600 lux caused a change of the potential of only ± 0.2 mV in the same solution [224]. An explanation is difficult, because the effect depends on equilibria in the solid phase, and the ionic conductivity in the membrane considerably exceeds the electronic conductivity [225–227]. In spite of a high degree of covalent bond character, leading to the low solubility in water, silver chloride can still be considered to be a predominantly ionic compound. The charge is carried chiefly by interstitial Ag$^+$ ions (Frenkel mechanism) and the conductivity in an almost perfect crystal may decrease to $10^{-9} \Omega^{-1}$ cm^{-1} [228]. However, conductivities of about $10^{-7} \Omega^{-1}$ cm^{-1} are commonly observed, owing to the effect of impurities and preparation procedures; electrolytically deposited AgCl has a conductivity one or two orders of magnitude higher [198]. Pressed mixed membranes have substantially higher conductivities than AgCl (e.g. *ca.* $10^{-4} \Omega^{-1}$ cm^{-1} for Ag$_2$S/AgCl [206]). The conductivities of Hg$_2$Cl$_2$/HgS membranes are substantially larger still [208].

4.2.1.2 SELECTIVITY

The effect of the pH is manifest only with lower chloride activities in alkaline solutions (e.g. for 3.5 mg/l. at pH > 6 [34, 208]). The determination is usually carried out at pH 2.0 – 4.5, where the ISE potential is most stable [229] and where the serious interference from CN^- and S^{2-} ions is suppressed to a considerable degree. Only trace concentrations of sulphide ($\leqq 10^{-7}M$) are permissible even at low pH; however, sulphide

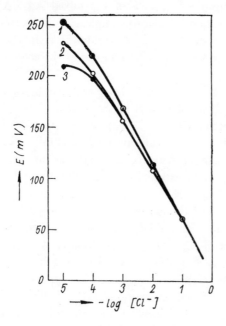

Fig. 4.4 Chloride ISE response to chloride and the effect of bromide in a binary mixture with chloride. $I = 0.1$, pH $= 3.7$. (1) — 0–1 × 10^{-6}M Br$^-$, (2) — 1 × 10^{-5}M Br$^-$, (3) — 1 × 10^{-4}M Br$^-$.

can be removed by oxidation, e.g. by a small amount of hydrogen peroxide, or by precipitation as PbS [11a]. The degree of interference in simple media can be assessed *a priori* on the basis of the solubility products and stability constants for compounds with silver [230] (the most important interferents are S^{2-}, CN^-, I^-, Br^-, $S_2O_3^{2-}$, CO_3^{2-}, NH_3, Ag^+, Cu^{2+}, Pb^{2+}, Tl^+ and Hg^{2+}), except for single-crystal AgCl-ISE's, where $k_{Cl,I}^{Pot}$ and $k_{Cl,Br}^{Pot}$ are substantially lower [201, 230a]. However, the composition of the measured solutions and the experimental conditions are important; theoretically, $k_{Cl,Br}^{Pot} = 2.0 \times 10^2$, but chloride may not be determinable even if present in a fiftyfold ratio to bromide [231, 232] (see Fig. 4.4.). The electrode surface is irreversibly poisoned in the presence of Br^-, I^- and S^{2-} (X), by precipitation of the silver salts, if

$$\frac{K_s(AgCl)}{[Cl^-]} \gtrsim \frac{K_s(AgX)}{[X]_1} \tag{4.5}$$

Silver sulphide can also be gradually formed on the Cl^--ISE membrane by decomposition of organic thio-compounds. On the other hand, prolonged exposure to a concentrated chloride solution leads to dissolution of AgCl from the surface of the mixed $Ag_2S/AgCl$ membrane, because of formation of chloro-complexes; the porous layer formed retards the electrode response. Chloride-ISE's damaged in these two ways can be regenerated only by removal of the corroded layer, e.g. by abrasion.

The response of a Cl^--ISE is faster and membrane corrosion slower when a fine AgCl suspension is added to the solution; in this way, the effect of interferents (Br^-, I^-, S^{2-}) can be suppressed when they are present at much lower concentrations than chloride. Higher concentrations of iodide can be removed by oxidation to iodine with $KMnO_4$ in an acidic medium, extraction of iodine into CCl_4 and removal of excess of $KMnO_4$ with H_2O_2; the oxidation can also be done with KNO_2 in $0.01M$ HNO_3 and the liberated iodine removed by boiling the solution [232]. Bromide and iodide can also be sorbed on a strongly basic anion-exchanger in the NO_3^--form and eluted with $0.5M$ KNO_3 [233].

The CN^-, NH_3 and $S_2O_3^{2-}$ species form stable complexes with Ag^+ and increase the solubility of AgCl and the content of Cl^- in the solution. The permissible concentration of $S_2O_3^{2-}$ is two orders of magnitude lower than that of chloride, while the concentration of NH_3 must not exceed that of chloride [34]. Cations forming poorly soluble chlorides also interfere. Concentrated Cu^{2+} solutions may reduce the Nernstian response [234].

The S^{2-}, I^- and Br^- ions seriously interfere with the response of some recently developed Hg_2Cl_2/HgS electrodes. The Orion 93-17 liquid ion-exchanger electrode tolerates small amounts of Br^- and I^- [83], but its other properties are poorer and it is used only in biochemistry [234a].

4.2.1.3 TITRATION OF CHLORIDE

Silver iodide- and sulphide-ISE's are more frequently employed in titrations of chloride, because they exhibit a lower detection limit and lower sensitivity to poisoning by bromide or iodide ions than Cl^--ISE's. These electrodes also allow successive titration of Cl^-, Br^- and I^- in the presence of each other, $Ba(NO_3)_2$ being added to the solution to suppress titrant adsorption on the silver halide precipitate [235]. The titration error has frequently been discussed in the literature [195, 197]; in the presence of bromide it depends on the $[Cl^-] : [Br^-]$ ratio. In precise chloride determinations it is better to separate bromide, e.g. by oxidation with H_2O_2 in dilute HNO_3 and reaction of the bromine with 8-hydroxy-quinoline [236] or by deposition of bromide on a strongly basic anion-exchanger [233]. Iodide does not interfere, as mixed precipitates of AgCl and AgI are not formed. Silver nitrate or mercuric perchlorate are used as titrants; the latter should not be added beyond the end-point, to prevent poisoning of the electrode by Hg^{2+} [129]. Titration of chloride is carried out in neutral or preferably weakly acidic solutions, possibly in the presence of 50 % alcohol or dioxan [237] or acetone [129], in order to obtain steeper titration curves. Titration in 75 % acetic acid has also been described [238]. In the presence of organic solvents the detection limit is decreased from *ca.* 5 mg/l. to tenths or hundredths of mg/l. Coulometric titration of chloride with electrogenerated Ag^+ ions has been performed with the Ag_2S electrode [239].

4.2.1.4 DETERMINATION OF CHLORIDE IN INORGANIC MATERIALS

Chloride-ISE's are chiefly employed in the determination of Cl^- in waters at a concentration below 1 mg/l., where visual titration does not yield good results [240]. In analyses of waters and soil extracts, a constant ionic strength of at least $0.1M$ is obtained with potassium or ammonium nitrate and the pH is adjusted to 2.3–3.0 with acetic or nitric acid [229, 232]. It is also possible to employ an acetate pH-buffer with a small amount of $Al(NO_3)_3$ to complex interfering organic compounds. Continuous monitoring of the Cl^--level in water has also been described [208a, 241].

Water-insoluble solid samples are dissolved in acids or decomposed by alkaline fusion, usually with sodium carbonate [242], or by sintering with a mixture of Na_2CO_3, KNO_3 and ZnO [243]. Phosphates can be dissolved in a mixture of perchloric acid and citric acid, triethanolamine added and the pH adjusted to 2.5 and the solution can be directly measured [123]. Silicates are best decomposed by pyrohydrolysis, where the danger of sample contamination by chloride is least and it is possible to determine chloride and fluoride simultaneously [229]. The same holds for the determination of Cl^- or HCl in gases and aerosols in continuous analysers [244].

4.2.1.5. DETERMINATION OF CHLORIDE IN ORGANIC MATERIALS

Liquid samples miscible with water are usually diluted with a KNO_3 solution to suppress the effect of the organic components; this is e.g. employed in analyses of body fluids. For the *in situ* determination of chloride in perspiration, in diagnosing cystic fibrosis, a specially modified ISE, Orion 417, can be employed [246, 246a].

Solid organic samples are usually extracted with water; however, when the overall chloride content (dissociated plus undissociated) is to be determined, the sample must be mineralized (e.g. according to Schöniger [137, 247] or by some of the methods described in Section 4.1.2). The Cl^- concentration is determined directly or argentometrically [237, 248]. In analyses of paper-mill pulping liquors, sulphide is first oxidized by H_2O_2, the peroxide is decomposed by boiling and the interferents are removed on a cation-exchanger in the H^+-form; Cl^- in the eluate is then titrated [249]. The titration determination has also been applied to molasses and sugar-mill products [250].

Elemental chlorine and hypochlorite in potable and industrial waters chlorinated with chloramine can be determined by using an iodide-ISE and the analyte-subtraction technique. The auxiliary solution, in which the loss of iodide is measured, contains e.g. $5.64 \times 10^{-5} M$ KI, $0.1M$ acetic acid, $0.1M$ sodium acetate and $10^{-3}M$ sodium citrate. Chlorine concentrations above 0.1 mg/l. can be determined [251].

4.2.2 Determination of Perchlorate

Chlorine compounds can also be determined in the form of chlorate [251a], but more often as perchlorate, by using a ClO_4^--ISE. Several electrodes, mostly LISE's with a relatively high selectivity, have been prepared. The best known electrode employs the formation of an ion-association complex between perchlorate and the Fe^{2+} complex with substituted 1,10-phenanthroline [17, 283, 284].

Perchlorates of substituted ammonium salts [218, 285], Brilliant Green [286], Methylene Blue [291] and organic radical salts, e.g. N-ethyl-benzothiazole-2,2'-azoviolene [287], can also be used. The detection limit of commercial ClO_4^--ISE's is 1–$2 \times 10^{-6} M$ (Orion, Corning), for an LISE with a radical salt $3 \times 10^{-6} M$, and higher for a solid-state electrode with the same pressed material [287].

The selectivity series is similar for all ClO_4^--ISE's: $ClO_4^- > SCN^- \sim I^- \sim BF_4^- > NO_3^- > Br^- > Cl^-$. Hence, voluminous and weakly hydrated anions interfere more seriously [288] ($k_{ClO_4^-, I^-}^{Pot} = 1 \times 10^{-2}$ [17, 284]). Chloride exhibits the smallest interference and is therefore used for the ionic strength adjustment. The Fe^{2+}-1,10-phenanthroline-ISE is sensitive to other large univalent anions, e.g. IO_4^- or MnO_4^- [289], ReO_4^- and SCN^- [290], and can be employed for their determination. An improvement in the selectivity was achieved by building the ion-exchanger into a PVC matrix [284, 285]; generally the electrode potential shifted to more positive values at extreme pH values. However, ClO_4^- can be determined even in $0.1 M$ HCl or $0.1 M$ NaOH (the Orion 92-81

Table 4.4

Survey of application of chloride-ISE's in the determination of chloride in inorganic and organic materials

Material	Ref.	Material	Ref.
water	232, 241, 252—254	benzoyl chloride hydrolysate	265
sea-water	255, 256	acetonitrile	266
soil extracts	252, 257—259	volatile organic compounds	267
boiler cleaning solutions	219	halogenated alkaloids	268
H_2SO_4 + NaCl solutions	83	glycosides	269
KOH	233	sugar-mill products	250
$Al(OH)_3$	260	grain syrup	270
plaster	83	tomato juice	83
metal-plating baths	261	plant tissues	271
Ca phosphates and		potato peel	251
halophosphates	219, 104	foodstuffs	272
silicate rocks	229	cheese	272, 273
paper-mill pulping liquors	249	butter and solidified fats	83
aerosols	123, 244	meat products	83, 274
pharmaceuticals	241, 262	ice-cream	83
pesticides	263	perspiration	275, 276, 276a
chloroisopropylate hydrolysate	264	urine	277, 278
		blood serum and plasma	277—282

ISE), provided that the acid or alkali concentration is identical in all solutions [292]. Direct potentiometry with the perchlorate electrode was employed in the determination of the solubility products of quarternary onium salts [293], $KClO_4$ and complex perchlorates of Co, Cu and Fe with ammonium, pyridine and 1,10-phenanthroline and in the study of the kinetics of the thermal decomposition of $HClO_4$ in aqueous solutions [294]. Perchlorate is usually determined by potentiometric titration with tetraphenylarsonium chloride at pH 4–7 [283, 289, 295, 296]. Excess of nitrate, chlorate, chromate or bromate distorts the titration curves. The titration is best performed at 2 °C ($10^{-4}M$ ClO_4^- is determined with an error of 1 % rel.). The titration has been applied to the determination of ClO_4^- in solid fuels [289]. Silver perchlorate has been employed for the titration of the hydrogen halides of quarternary onium salts [293]. Perrhenates can be titrated with $AgNO_3$ in an acetate buffer of pH 5.5, employing the ClO_4^--ISE [290].

4.3 DETERMINATION OF BROMINE COMPOUNDS

4.3.1 Determination of Bromide

4.3.1.1 BROMIDE-SELECTIVE ELECTRODE

Sparingly soluble silver and mercurous chlorides and bromides have similar properties and therefore solid-state chloride- and bromide-ISE's are also similar. Homogeneous Br^--ISE's have membranes pressed from an $Ag_2S/AgBr$ [206] or HgS/Hg_2Br_2 [297, 297a] mixture or made from an AgBr single-crystal [34, 202]. Heterogeneous electrodes containing silicone rubber [211, 212] or other materials [213] can also be made. Bromide-LISE's [218, 298 – 300] and microelectrodes [220] have been designed.

The lower solubility of silver bromide than of the chloride ($K_s(AgBr) = 7.7 \times 10^{-13}$, $pK_s = 12.11$ [221]) permits a lower detection limit, *ca.* $1 \times 10^{-6}M$ Br^-, i.e. 80 µg/l. in simple solutions. An electrode based on a pressed mixture of HgS/Hg_2Br_2 [297] exhibits a detection limit one order of magnitude lower; this ISE has better selectivity for bromide with respect to chloride, but worse with respect to iodide, compared with the AgBr-based electrode. According to various authors [34, 231, 232], the AgBr electrode tolerates a 10–200-fold ratio of chloride; the theoretical k_{Br^-, Cl^-}^{Pot} is 4.9×10^{-3}. Adsorption of chloride on the membrane causes reversible poisoning of the Br^--ISE, which is removed by immersion in a $10^{-3}M$ Br^- solution [232]. Larger amounts of iodide interfere (theoretical $k^{Pot}_{Br^-, I^-} = 5.1 \times 10^3$), although its effect decreases in the presence of citrate buffers of pH 2.5. Iodide can be partly removed by oxidation with H_2O_2 at laboratory temperature [231]. Irreversible poisoning on exposure to concentrated iodide solutions can be removed only by polishing the membrane and even then the original electrode properties need not necessarily be attained [34].

Other ions interfering with the chloride-ISE also disturb the determination of bromide. Hydroxide interference starts at pH > 6 (for a Br^- concentration of 0.80 mg/l.). Thiosulphate interferes at concentrations higher than one tenth of the bromide concentration and ammonia when in hundredfold ratio to bromide [34].

4.3.1.2 TITRATION OF BROMIDE

Bromide is titrated argento- or mercurimetrically with the Br^--ISE, or, more frequently, the AgI- or Ag_2S-ISE's, similarly to the titration of chloride (see Section 4.2.1.3). Problems are encountered in titration of Br^- in the presence of Cl^-, owing to formation of a mixed precipitate and consequent positive error, and to sluggish AgBr electrode-potential stabilization. The error can be decreased by using the Gran titration technique. A similar problem is experienced in titration of bromide in the presence of thiocyanate; a separation procedure is proposed [301] for SCN^- (using CuSCN) and the Br^- concentration is calculated from the difference of $AgNO_3$ consumptions before and after SCN^- separation.

In bromide and chloride mixtures, bromide can also be determined from the difference in the $AgNO_3$ consumptions before and after bromide separation either by oxidation with H_2O_2 and reaction with 8-hydroxyquinoline [236], or oxidation with HNO_3, the bromine being expelled by passage of air through the reaction mixture [302]. The two halides can also be separated on a strongly basic anion-exchanger in the NO_3^--form [233]. Bromide concentrations above 1 mg/l. can be titrated.

4.3.1.3 DETERMINATION OF BROMIDE IN NATURAL MATERIALS

Preliminary separations are frequently required. Iodide is separated by oxidation; sulphide, thiocyanate, thiosulphate and cyanide are also removed in this way. The most serious difficulty is the frequent presence of excess of chloride. For example, in surface, spring and potable waters, in which the bromide concentration is between 0.08 and 0.8 mg/l., the determination is successful only if the chloride concentration does not exceed 5 mg/l.; otherwise bromide should be separated (see Section 4.3.1.2).

Positive errors and potential drift are also encountered in the determination of bromide in body fluids [279, 303]; a positive error in blood serum and plasma analysis can be caused not only by chloride, but also by cysteine. A determination of bromide in urine has also been described [303a]. The Br^--ISE has also been used for the determination of bromide in atmospheric precipitation [78].

Bromine and hypobromite can be determined in aqueous media by the analyte-subtraction method, by measuring the residual iodide concentration with the AgI-ISE [251].

Potentiometric titration is employed more frequently than direct potentiometry. For example, bromide is determined in non-alcoholic beverages argentometrically [305] and in combustion products of volatile organic compounds mercurimetrically [271].

It is found that the electrode response is the same in non-aqueous media (alcohols, acetic acid) as in water [304].

4.4 DETERMINATION OF IODINE COMPOUNDS

4.4.1 Determination of Iodide

4.4.1.1 IODIDE-SELECTIVE ELECTRODES

Four types of electrodes have been constructed, namely, homogeneous electrodes with pressed Ag_2S/AgI membranes [17, 206, 220, 306] or HgS/Hg_2I_2 membranes [297], homogeneous membranes containing AgI alone [34, 202], heterogeneous electrodes with AgI in silicone rubber [212, 307, 308] or other plastics [213] and LISE's with various electroactive substances [218, 309 – 312].

Heterogeneous iodide electrodes were important in the early history of ISE's [313], but later were superseded by homogeneous electrodes with faster response and longer life-time [314, 315]. Preparation of homogeneous electrodes from AgI alone (e.g. Crytur and Corning electrodes) is complicated by the polymorphism of AgI. The system is very complicated under extreme conditions [316], but at laboratory temperature only the β-AgI (hexagonal) and γ-AgI (cubic) forms exist. On precipitation from a solution with excess of Ag^+, mostly γ-AgI is obtained; with excess of iodide, predominantly β-AgI precipitates [317]. Membranes with γ-AgI are more sensitive to iodide [318], but a slow modification occurs on contact with iodide solutions, accompanied by volume changes in the membrane and a decrease in the electrode sensitivity. Modifications also occur during pressing and on exceeding 146 °C (for the phase diagram see [319]).

On the other hand, pressed Ag_2S + AgI membranes containing a certain amount of Ag_3SI [17, 318] do not change with time; however, changes in the E vs. a relation on contact with solutions still occur. The conductivity of these membranes is $ca.$ $10^{-4} \Omega^{-1}$ cm^{-1} [318], while that of polycrystalline pellets probably depends on the contact conductivity [320].

The AgI solubility product, $K_s(AgI) = 1.5 \times 10^{-16}$, $pK_s = 15.82$ [221] [somewhat lower values have also been reported, e.g. $K_s(AgI) = 8.6 \times 10^{-17}$], yields a theoretical detection limit of $ca.$ $10^{-8} M$ I^-. However, reproducibility is poor for pI > 5 in unbuffered solutions [207, 318]. A slope higher than Nernstian, observed with an AgI/AgBr/Ag_2S mixture [207], appears frequently with the Orion 94 − 53A electrode (an Ag_2S/AgI mixture) and can also occur with freshly prepared γ-AgI electrodes [318]. Slopes lower than Nernstian appear commonly at pI > 5, for example, after electrode exposure to a concentrated iodide solution. The E vs. a dependence is affected by the method of electrode preparation (Ag^+ or I^- excess) (see Fig. 4.5) and by the memory effect (see Chapter 3). The immediate vicinity of the electrode surface can contain excess of silver ion or of iodide because of adsorption effects, leading to deviations from the Nernstian slope and to changes in the potential on cessation of stirring [318]. These effects of the electrode history and possible solution oxidation [322, 322a] cause poor reproducibility of measurements at pI > 5. An assumption that these effects are also caused by defects in the solid phase [321] now seems improbable [322a].

The solid-phase effects apparently also determine the photosensitivity of silver halide electrodes [323 − 325]; common changes in light cause relatively small potential changes of I^--ISE's, but these depend on the Ag^+ concentration in the solution [318].

It is possible to measure at pH 1–12 with the I^--ISE, but pH 2.5–3.0 is preferable for low iodide activities [231, 326], to prevent OH^--interference and AgI reduction in alkaline solutions [327]. In the presence of HI, a sulphur-containing layer can be formed on the surface of the

Ag$_2$S/AgI-ISE [327a]. Choice of the external reference electrode is also important [231]. Though higher concentrations of bromide (more than $10^{-2}M$; theor. $k_{I^-, Br^-}^{Pot} = 1.9 \times 10^{-4}$) and even iodide [315, 328, 328a] (formation of complex AgI$_3^{2-}$) cause temporary electrode poisoning,

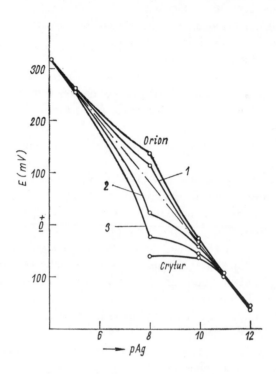

Fig. 4.5 Deviations from the Nernstian dependence at pAg = 5–11 for various iodide ISE's. A 0.1M NaNO$_3$ solution was used for pAg = 8. (1) — AgI membrane containing predominantly γ-AgI; pressure 5.07×10^8 Pa; (2) — AgI membrane containing predominantly β-AgI; pressure 5.07×10^8 Pa; (3) — Ag$_2$S + AgI membrane, pressure 2.53×10^8 Pa. The electrode surface was renewed by polishing. The E vs. pAg curves do not exhibit a maximum or minimum at pAg = 8 in practice, but it is difficult to prepare unbuffered Ag$^+$-solutions in this activity region.

chloride does not interfere even at high concentrations (theor. k_{I^-, Cl^-}^{Pot} = 9.6×10^{-7}). The degree of interference from species complexing Ag$^+$ (e.g. S$_2$O$_3^{2-}$ or NH$_3$) is lower than with chloride and bromide electrodes, owing to the lower solubility of AgI. In the alkaline region CN$^-$ interferes, at concentrations comparable with that of iodide, and so, of course, does sulphide, which may be present only at trace concentrations ($\leq 10^{-7}M$) at any pH value. The interference from CN$^-$, on which its determination is based (see Section 4.5), can be suppressed by adjusting the pH to 2–3 [326, 329]. Strong oxidants can oxidize iodide to iodine in the membrane, while strong reductants form metallic silver in alkaline media [327].

4.4.1.2 DETERMINATION OF IODIDE
IN VARIOUS MATERIALS

Argentometric and mercurimetric titrations permit the determination of iodide down to 1 mg/l. in neutral or weakly acidic media, even in the presence of SCN^-, CN^-, Br^- and Cl^- [248, 330]. High concentrations of Br^-, SCN^- and $S_2O_3^{2-}$ cause an error in the mercurimetric determination of iodide, but chloride causes only a decrease in the potential break; cyanide is co-titrated and must be removed in the form of HCN, by boiling the acidified solution.

The iodide-ISE is successfully applied to determinations in all kinds of waters, even in the presence of rather high concentrations of chloride and bromide. Bromide slows down the I^--ISE response and its concentration must not exceed 5 g/l. [329]. When iodide is to be determined in the presence of other iodine compounds, the test solution is brought to pH 4.5 and KNO_3 is added to a final concentration of $0.1M$. If, however, total iodine is to be determined in waters, all higher oxidation states are reduced to I^- either by sulphite or better with ascorbic acid [329]. Iodine in iodized sodium chloride and in other salts is determined similarly.

In the determination of iodine in the atmosphere the air is passed through a trap containing $1M$ NaOH; the solution is then acidified with HCl and the procedure for the determination in waters is followed [329]. Mineral materials are decomposed by sintering with a mixture of $NaKCO_3$ and MgO, the melt is extracted with water, the extract is filtered and the procedure above is employed for the determination of iodide. Iodate and iodide in the presence of one another are determined from two measurements, before and after reduction of the IO_3^- by Al-foil in alkaline solution [331]. Selenium samples are dissolved in or fused with NaOH for the determination of iodine. The solution is then acidified and the standard addition method is employed for the determination [112]. The specific activity of ^{131}I and ^{125}I has also been determined with the help of an I^--ISE [332].

The determination of iodine in organic materials is similar to that of the other halides, i.e. for the determination of the total iodine the sample must be mineralized and all iodine reduced to iodide either by Al-foil in

alkaline medium [333] or by sulphurous or ascorbic acids in acidic
solution. The iodide-ISE responds to iodide even in polar organic and
mixed solvents [304]; this can be utilized for the determination of iodide
in organic materials soluble in aliphatic alcohols or in glacial acetic
acid.

Table 4.5

Survey of application of iodide-ISE's in the determination of iodide and iodine
compounds in natural materials

Material	Ref.	Material	Ref.
waters	329, 330	determination of IO_3^-	331
inorganic salts	329, 334	determination of IO_4^-	337
mineral raw materials	329	thyroxine	333
selenium	112	serum	338
during reaction of H_2O_2		bile	331
with IO_3^-	335	milk	339
during IO_3^- reduction with		pharmaceuticals	333
hydrazine	336	plant material	340

4.5 DETERMINATION OF CYANO COMPOUNDS

4.5.1 Determination of Cyanide

Several solid-state ISE's can be employed for the determination of
cyanide. The cationic (Ag^+) glass electrode permits only titration of
cyanide (see e.g. [341]), but the I^-/CN^- membrane electrodes [326,
342], an Ag/AgI electrode of the second kind [342a] and Ag_2S
electrode [343 − 345a] are also suitable for direct potentiometry. The response principle of the most common I^-/CN^- electrode is given by the
reactiȯn

$$AgI(s) + 2 CN_2^- \rightleftharpoons Ag(CN)_2^- + I^-; \quad K \sim 10^4 \quad (4.6)$$

because

$$\beta_2 = \frac{[Ag(CN)_2^-]}{[Ag^+][CN^-]^2} = 2.7 \times 10^{20} \quad (4.7)$$

gives an a_{Ag^+} lower than that given by the AgI solubility product. The
measurement is always carried out in stirred solutions to enhance the
transport of cyanide and iodide and the electrode potential is virtually

determined by the activity of the iodide displaced from the electrode surface (or by a_{Ag^+} through a_{I^-} and cyanide). The response is in principle identical for AgI and AgI/Ag$_2$S membranes, but with the latter the porous Ag$_2$S layer retards diffusion and consequently the electrode response is slower. The experimental selectivity coefficients, k_{I^-, CN^-}^{Pot}, differ only very little for the two membranes (1.6–1.7 for AgI [347] and 1.3–1.5 for AgI/Ag$_2$S [348]) and are lower than the value of 2 expected from the stoichiometry of Eq. (4.6).

Though I^-/CN^- electrodes should not be used at CN^- concentrations higher than $10^{-3}M$ because of rapid dissolution of AgI, the Ag$_2$S electrode can be employed in concentrated cyanide solutions. With $[CN^-] > 10^{-2}M$, Ag(CN)$_3^{2-}$ is formed:

$$6 \, CN^- + Ag_2S \quad \rightarrow \quad S^{2-} + 2 \, Ag(CN)_3^{2-};$$
$$\beta_3 = 9.55 \times 10^{21} \, [346]$$

(4.8)

The experimental $k_{S^{2-}, CN^-}^{Pot} = 6.3 \times 10^{-3}$ in $1M$ KCN [344]. At cyanide concentrations less than $10^{-2}M$, the response slope increases to more than 100 mV/pCN and the potential-reproducibility becomes poorer with decreasing cyanide concentration (although it is still better than with a silver metal electrode), because the sulphide concentration formed is insufficient for potential-control and the amount of Ag(CN)$_2^-$ is too small to buffer the Ag$^+$ activity. Measurement of Ag$^+$ activity with an Ag$_2$S electrode is the basis of the analyte-subtraction method for the determination of cyanide [343, 345, 349, 350]. The sample is added to a buffered solution with a sufficient concentration of K[Ag(CN)$_2$], containing *ca.* 1 % excess of cyanide to prevent precipitation of AgCN. On addition of a cyanide sample solution, a_{Ag^+} decreases according to Eq. (4.7). If the [Ag(CN)$_2$]$^-$ concentration is selected so that it can be considered constant during the measurement, then a tenfold increase in the cyanide activity decreases a_{Ag^+} by a factor of 100, yielding a slope of 118.3 mV/pCN compared with 59.2 mV/pCN for the I^-/CN^- electrode at 25 °C. For example, the detection limit is 0.03 μg/l. ($10^{-6}M$ CN^-) for [Ag(CN)$_2^-$] = $10^{-5}M$. Moreover, the use of unstable standard cyanide solutions is avoided and a 1000-fold excess of NH$_3$ [345, 349] and probably also halides will not interfere. The same technique is also employed with the gas-sensing HCN electrode [351], with a detection limit of *ca.* $10^{-7}M$.

Cyanide is present as the ion CN^- or as the dissolved gas HCN in the solution, in dependence on the pH. The pK_a value is 9.31 [221], i.e. only at pH > 11.3 is more than 99 % of the cyanide present as CN^-. As interference of OH$^-$ ions [326, 345] and reduction of AgI [329] can occur at pH > 12, a pH-range of 11.3–12.0 is used for the determination. A detection limit of 2–5 ppM* (ca. $10^{-7}M$ CN^-) was found for pH 11.5 with an I^-/CN^--ISE [326].

There is controversy in the literature concerning the sensitivity of the AgI electrode to the poorly dissociated HCN. On the one hand, a better agreement with experiment is attained by including the reaction

$$AgI + HCN \rightleftharpoons Ag(CN)_2^- + 2H^+ + I^- \qquad (4.9)$$

in the response scheme in the absence of interfering anions [351]; on the other hand, changes in the potential with changing pH (at pH < 10) indicate that a_{CN^-} varies in a way that excludes electrode sensitivity to HCN [342].

The determination of cyanide is further interesting because not only free CN^-, but also all cyanide complexed in compounds that are less stable than $Ag(CN)_2^-$ (for which β_2 is ca. 10^{20} [342, 348, 353]), is detected, e.g. $Cd(CN)_4^{2-}$ or $Zn(CN)_4^{2-}$. Stronger cyanide complexes must be decomposed, e.g. by acidification with acetic acid to pH 4 and heating to 50 °C, possibly in combination with a displacement reaction with EDTA [350]. This procedure, however, has been criticized [326] on the grounds that the complex decomposition is incomplete and CN^- may volatilize as HCN during heating of acidic solutions. A more elegant decomposition method is by ultraviolet-irradiation in a quartz tube [354]. Sulphide is added to the sample, acidified with phosphoric acid, in order to bind the liberated metals. The excess of sulphide is removed by addition of Bi^{3+} which does not interfere in the determination of total cyanide (above 10 μg/l.) at pH 11.5 with the I^-/CN^- electrode [326].

The selectivity is similar to that for the determination of iodide, sulphide and silver with these electrodes. The interference from sulphide is suppressed by addition of Bi^{3+}, and iodide is measured with an identical I^-/CN^--ISE at pH 2–3 and its value is subtracted from the value obtained in alkaline medium, the sensitivities for cyanide and iodide being taken as identical [326, 329]. ISE's for the determination of cyanide are stored in distilled water, free of cyanide, or in the air. The electrode should not be exposed to cyanide concentrations more than two orders of magnitude higher than the minimum concentration to be determined, since otherwise a following measurement of a low concentration is sluggish [343]. Cyanide can also be titrated with $AgNO_3$ or $Hg(ClO_4)_2$ in neutral or weakly alkaline solutions (pH 9), the AgI or better the Ag_2S electrode being used [342, 343] (Fig. 4.6). The titration curves are

---

*Parts per milliard (10^9).

symmetrical and steep, permitting the determination of *ca.* $10^{-4}M$ CN$^-$ (above 3 mg/l.) with an error of about $\pm 0.1\%$ rel. High concentrations of Br$^-$, SCN$^-$ and S$_2$O$_3^{2-}$ interfere in the mercurimetric determination; excess of Cl$^-$ merely makes the potential break more drawn-out [329].

The determination of cyanide in waters, especially waste waters, is important [247, 326, 329]. Before the measurement, KNO$_3$ is added to a concentration of *ca.* 0.1M and the solution is made alkaline to thymolphthalein indicator. Free cyanide can be determined in a range of 0.020 –260 mg/l.; low concentrations are better determined by Gran titration, adding Ag(CN)$_2^-$ in excess and measuring the Ag$^+$ activity to avoid preparation of unstable dilute standard CN$^-$ solutions.

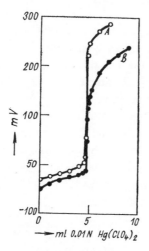

Fig. 4.6 Mercurimetric titration of $5 \times 10^{-4}M$ CN$^-$ (A) and in equimolar mixture with chloride, (B), using the Crytur iodide-ISE (reprinted from ref. [329] by permission of Academia).

For precise direct determinations below 0.3 mg/l., preliminary distillation of HCN and its absorption in a KOH solution is sometimes recommended [355]. Flow-through continuous analysers have been constructed [356, 357], e.g. the Orion Model 1206 [358]. Cyanide can be determined in plating baths [261, 359], in photographic developers [360] (adding EDTA, NTA and citric acid to prevent adsorption of Ag$^+$ on the electrode surface) and in 0.01M hexacyanochromate(III) (in 0.1M Na$_2$HPO$_4$) in a concentration range of 3–26 mg/l. AgI is suspended in the solution to achieve better reproducibility and faster electrode response [361].

The I$^-$/CN$^-$-ISE has been applied to the determination of amygdalin after its enzymatic hydrolysis [269, 368, 368a]. The procedure was employed in analyses for cyanoglycosides in Sudan grass [362], fodders [363], alcoholic beverages [364] and in various biological materials [365]. In analyses of aerosols, e.g. cigarette smoke, the sample is passed through an "Ascarite" filter, which is then extracted with NaOH solution containing $Pb(NO_3)_2$ [366]. Cyanocobalamin is determined in pharmaceuticals after liberation of HCN by reduction with $SnCl_2$ or calcium hypophosphite and absorption in an alkaline solution [367]. Procedures have been developed for continuous determination of CN^- [367a] and for determination of CN^- in steelwork effluents [367b].

4.5.2 Determination of Thiocyanate

Solid-state thiocyanate-ISE's contain AgSCN [369, 370] or $Hg_2(SCN)_2$ [297]. Crystal Violet, tetraphenylarsonium [371] or Aliquat 336S [218] function as cations in LISE's applicable to the determination of thiocyanate. Commercial perchlorate (Orion 92−81) [290] and bromide [372] electrodes are also sensitive to SCN^- ions.

The cyanide-ISE [373] can also be employed for the determination of thiocyanate, by conversion of the latter into BrCN with bromine water and subsequent reduction by sulphite. This determination is not disturbed by excess of cyanide, which can be expelled by boiling the solution at pH 8. Interfering cations (e.g. Cu^{2+}, Cd^{2+}, Fe^{3+}, Cr^{3+}) can be removed with a cation-exchanger [373].

The solubility of silver thiocyanate is between that of silver chloride and of bromide ($K_s(AgSCN) = 1.16 \times 10^{-12}$, $pK_s = 11.9$ at 25 °C [221]). The detection limit is $5 \times 10^{-6}M$ (0.30 mg/l.) for the AgSCN electrode [83, 297] and $5 \times 10^{-7}M$ (30 μg/l.) for the $Hg_2(SCN)_2$ electrode [297]. The main interfering ions are identical with those for the Cl^--ISE. Some LISE's are relatively selective for SCN^- and I^- with respect to Cl^- and Br^- [371], which interfere with solid-state electrodes when present at higher concentrations. LISE's exhibit a broader useful pH-range than solid-state ISE's.

The detection limit of solid-state ISE's is decreased when mixed media are used (containing e.g. ethanol, dioxan, acetone or dimethylformamide). For example, $2 \times 10^{-5}M$ SCN^- can be titrated in a 4:1 mixture of dioxan with water [374]. In non-aqueous media of methanol, acetone or acetonitrile, the SCN^- activity can be measured directly or titrated [370] with Ni^{2+}, Co^{2+}, Cd^{2+}, Hg_2^{2+} or Hg^{2+} nitrate.

Using the Ag_2S-ISE, it is possible to titrate thiocyanate argentometrically in the presence of bromide. Thiocyanate can be separated from bromide by precipitation as CuSCN in an ammoniacal medium at pH 9. The precipitate is dissolved in dilute HNO_3 [301]. Thiocyanate can also be titrated mercurimetrically at pH 3.5–4.0, the AgI-ISE being used [329]. An AgSCN-ISE was used for the study of thiocyanate complexes [374a].

4.6 DETERMINATION OF SULPHUR COMPOUNDS

4.6.1 Determination of Sulphide

Sulphide is determined only with the ISE's based on Ag_2S, which can be prepared by vacuum pressing [17, 375] or by dispersion of the Ag_2S in silicone rubber [376] or polyethylene [377]. Single-crystals are not suitable because of non-stoichiometry (excess of silver) introduced during crystallization.

The reaction mechanism is based on variations in a_{Ag^+} at the electrode:

$$a_{Ag^+} = \sqrt{\frac{K_s(Ag_2S)}{a_{S^{2-}}}} \qquad (4.10)$$

Hence, the electrode response slope is 28.6 mV/pS^{2-} at 25 °C. The sulphide activity is pH-dependent:

$$S^{2-} + H_2O \; \rightleftharpoons \; HS^- + OH^- \qquad (4.11)$$

$$HS^- + H_2O \; \rightleftharpoons \; H_2S + OH^- \qquad (4.12)$$

$$a_{S^{2-}} = \frac{\gamma_{S^{2-}}[S]_{total}}{1 + \dfrac{a_{H^+}}{K_{a2}} + \dfrac{(a_{H^+})^2}{K_{a1}K_{a2}}} \qquad (4.13)$$

Consequently, the electrode potential becomes more positive with decreasing pH, as has been verified experimentally [376, 378]. The pK_{a1} value is 7.04 [221], the pK_{a2} value is still uncertain but in the range 12–14; a recently reported pK_{a2} value [379], 13.88 ± 0.02 (25 °C) indicates that S^{2-} will predominate only at pH > 13.9. The determination is therefore carried out in $1M$ alkali; it is possible to determine sulphide at lower pH, but the pH must simultaneously be measured and a correction made by using Eq. (4.13).

The gas-sensing H_2S-electrode [351, 351a] is based on this principle (the gas diffuses into a citrate buffer of pH ~ 5 on the surface of an Ag_2S electrode). Hydrogen sulphide can be determined at concentrations of 10^{-2}–$10^{-6}M$ [83].

The very low solubility of Ag_2S ($K_s(Ag_2S) = 1.6 \times 10^{-49}$ at 18 °C [221]; from ISE measurements, $K_s(Ag_2S) = 1.48 \times 10^{-51}$ [378] and 4.4×10^{-51} if $pK_{a2} = 13.9$ or 3.3×10^{-50} if $pK_{a2} = 12.9$ [201]) indicates a very wide accessible activity range of the electrode, verified by measurements in buffered solutions [10, 344]. On the other hand, S^{2-} down to $3 \times 10^{-6}M$ (10 µg/l.), or *ca.* $10^{-7}M$ when metals are removed from the solution [381], can be determined in unbuffered solutions in which sulphide is protected against oxidation. Still, sulphide must sometimes be preconcentrated for its determination in natural waters, either by distillation of H_2S into an alkaline absorption solution [380] or by co-precipitation on ZnS or $Zn(OH)_2$ [381]. If concentrations below $10^{-5}M$ are to be measured, a 1-min exposure of the electrode to a solution with a pS^{2-} of about 4 is recommended in order to shorten the response time. In view of easy oxidation of S^{2-} and Ag_2S, the solutions measured must be deaerated by passage of an inert gas, or, better, by addition of ascorbic acid to the solution [322, 382], possibly in combination with sodium sulphite [380]. Electrode surfaces damaged by oxidation or the effect of polysulphides or aromatic polyhydroxy compounds must be renewed by grinding and polishing.

What was said about the iodide-ISE concerning poor potential reproducibility at pX > 5 holds to even greater degree for the Ag_2S electrode; owing to common deviations of Ag_2S from stoichiometry and to oxidation, a tendency to supersensitivity is generally observed at pS^{2-} > 5 [321, 322] which decreases with decreasing silver content in the membrane [34]. Especially large deviations, starting at pS^{2-} > 4, were found with single-crystal electrodes; the deviations decreased after tempering of the crystals in sulphur vapours [34]. The effect of the preparation procedure was also observed earlier [344, 377, 378].

Solid internal contact electrodes are at present most often employed. In contrast to silver halides, direct contact of silver or another metal with Ag_2S is unsatisfactory (positive potential deviations up to $+45$ mV at $a_{Ag^+} = 1$ [323, 344, 383] and poorer reproducibility). Therefore an intermediary layer of AgI is employed ($Ag/AgI/Ag_2S$).

The selectivity of the Ag_2S electrode is exceptionally high; only higher concentrations of cyanide (above $10^{-3}M$) and, of course, metals precipitating sulphide, interfere. The effect of light is very small, although temporary potential-variations of less than 2 mV have been observed on change in the light intensity [344].

The Ag_2S-ISE is used more frequently for titrations of sulphide, halides, cyanide, thiocyanate and SH-compounds than for direct potentiometry. As titrants, nitrates or perchlorates of Ag^+, Hg_2^{2+} and Cd^{2+} are employed. Argentometric sulphide titration [378], yielding potential breaks of up to 1100 mV, is best performed in $0.1M$ NaOH and $0.1M$ NH$_3$ [384], with simultaneous deaeration. Titrations of S^{2-} with Hg^{2+}, Pb^{2+} or sodium plumbate [385] are carried out in $1M$ NaOH, with possible addition of an anti-oxidation buffer ($0.4M$ ascorbic acid, $2M$ NaOH and $2M$ sodium salicylate) [83] or of hydrazine saturated with PbS [386]. Titration with a lead(II) salt permits the determination of sulphide in the presence of Cl$^-$, Br$^-$, I$^-$, SO$_3^{2-}$ and SCN$^-$ [385]. The Ag$_2$S electrode also responds to Hg$_2^{2+}$ and Hg^{2+}; therefore thiosulphate in the presence of sulphide and phenylmercaptans, and polysulphides after conversion into thiosulphate by sodium sulphite, can be titrated mercurimetrically [387, 388]. A mixture of sulphide and polysulphides can be titrated argentometrically after conversion of the polysulphides into thiocyanate by KCN [389].

Direct potentiometric determination of S^{2-} and H$_2$S in waters is frequently carried out; the samples are conserved by addition of 25 ml of a $0.5M$ ascorbic acid, $10M$ NaOH and $0.1M$ salicylic acid mixture to 1000 ml of water. Before the analysis, 50–100 ml of the conserved water are mixed with 10 ml of a buffer comprised of $0.4M$ ascorbic acid, $10M$ NaOH and $0.05M$ Na$_2$SO$_3$ [380]. Sulphides in soil extracts [390] and river sediments [391] are determined similarly; the latter analysis can also be effected by Gran titration with Cd(NO$_3$)$_2$. Argentometric and mercurimetric titrations have also been automated [392, 393]. With mineral raw materials, e.g. silicates [394], technical iron [395], etc., in which sulphur is contained in the elemental form or as insoluble sulphides, the samples must be oxidatively decomposed with acids. Soluble sulphates formed are then reduced in solution, e.g. by hypophosphorous acid with catalysis by HI. The H$_2$S formed is distilled into an alkaline solution of an anti-oxidation buffer and S^{2-} is determined. Hydrogen sulphide from aerosols and gaseous mixtures (e.g. the atmosphere [10]) or from cigarette smoke [396] is absorbed and measured analogously. Sulphur in mineral and organic materials can be converted into H$_2$S by hydrogenation in a platinum tube at 900 °C [380]. The H$_2$S is again absorbed in a buffer or in a $1M$ KOH/$1.5M$ NH$_2$OH mixture [397]. A procedure for converting sulphur into H$_2$S by pyrolysis in NH$_3$ at 800 °C in a quartz tube has been described. The H$_2$S is absorbed in $1M$ NaOH containing 5 % of glycerine and S^{2-} is titrated with an ammoniacal solution of AgNO$_3$ [398]. For analyses of black liquors, the analyte-subtraction method with an AgNO$_3$ standard solution is proposed. Polysulphides are converted into sulphide by Na$_2$SO$_3$ [399]; the Cd^{2+}-ISE is sometimes more suitable for this analysis, as reduction of Ag^+ by polysulphides and aromatic polyhydroxy compounds is avoided [400]. The S^{2-}-ISE is further employed for the study of hydrolysis of sulpho-derivatives and applied for titration of thiols in petroleum products [401] and disulphides in proteins [402]. Thiols [403] and phenylmercaptans [387] are mercurimetrically titrated in the presence of S^{2-}, $S_2O_3^{2-}$ and polysulphides. Thiourea [404] and thioacetamide (in $1M$ NaOH or $1M$ NaOH $+ 0.1M$ NH$_3$) as well as some thiourea derivatives, such as uracil and 2,4-dithiouracil (argentometrically in an acetate buffer of pH 5.6 [405]) can be analogously titrated. Mercaptobenzothiazole can be titrated in 80 % acetone [377].

Table 4.6

Application of sulphide-ISE's to inorganic and organic materials

Material	Ref.	Material	Ref.
waters	377, 380, 381, 406	polysulphides	385, 386, 388, 389
sea-water	392, 407, 408	organic compounds	397, 413, 414
waste waters	406, 409	thiols	398, 401, 403, 415
soil extracts	390	disulphides in proteins	416
river mud	391	thiourea	404
atmosphere	10	thioacetamide	404, 414
cigarette smoke	396	ammonium diethyldithio-	
beer	410	phosphate in flotation fluids	415
paper-mill pulping liquors	388, 364, 400	phenyl mercaptan	387
		uracil and 2,4-dithiouracil	405
in paper-mill and leather industry	411, 412	mercaptobenzothiazole⁻	377

4.6.2 Determination of Sulphate

Difficulties in the determination of sulphate, and its analytical importance, have stimulated efforts to develop an SO_4^{2-}-ISE. However, mineral baryte [417] and heterogeneous BaSO$_4$-containing membranes [418, 419, 419a] exhibit a poor selectivity toward univalent ions. Older types of ISE's reversible to SO_4^{2-} [420] often contained PbSO$_4$, which is also the main component of the SO_4^{2-}-ISE [421], containing 32 mole % PbS, 31 mole % Ag$_2$S, 32 mole % PbSO$_4$ and 5 mole % Cu$_2$S. The detection limit of this electrode is $10^{-5}M$ SO_4^{2-} and it can be used from pH 3 to 8.5; I^-, HPO_4^{2-} and SO_3^{2-} interfere.

The electrode must be frequently conditioned and its surface cleaned; this is probably the reason why it is not manufactured commercially. Therefore, sulphate is determined indirectly, mostly titrimetrically. For the end-point detection, the temporary response of Na^+-sensitive glass ISE's to Ba^{2+} can be used [422]; further, the Ba^{2+}-ISE [423, 423a] can be employed. For sulphate titration with barium an Fe^{3+}-ISE can be used to monitor Fe^{3+} ions liberated from the complex with SO_4^{2-} [424] or the non-Nernstian response of the Pb^{2+}-ISE to SO_4^{2-} can be used [425]. However, titration of sulphate with Pb^{2+} in a mixed solvent, using the Pb^{2+}-ISE, is most frequently employed [426] ($K_s(PbSO_4)$ $= 1.06 \times 10^{-8}$ at 18 °C in H_2O [221]; 70 % methanol is more suitable than 50 or 60 % dioxan [427, 428], as dioxan is unstable and gradually decomposes to H_2O_2, which oxidizes the electrode surface). In pure sulphate solutions, more than 1 mg/l. can be titrated with an error of 0.2–1 %; in natural materials the detection limit is one order of magnitude higher. Anions forming less soluble lead salts than $PbSO_4$ interfere, as does phosphate, which must be removed from the solution by ion-exchange or its effect suppressed by addition of $1M$ KNO_3 + $0.1M$ $La(NO_3)_3$. Chloride, nitrate and bicarbonate interfere seriously only when present in 100-fold amount relative to sulphate [426]. The effect of HCO_3^- can be suppressed by bringing the solution to pH 4. Fluoride can be masked by H_3BO_3; however, if chloride is present simultaneously (danger of formation of a PbClF precipitate), it is better to expel fluoride in the form of HF by boiling the solution after acidification with perchloric acid.

Titration of sulphate, using a Pb^{2+}-ISE, has been applied to soil extracts [427] and sea and mineral waters [429]. In these strongly mineralized waters, chloride and bicarbonate are separated by use of cation-exchangers in the Ag^+- and H^+-forms. Sulphate can also be determined in nickel-plating baths, by using the analyte-subtraction method with a standard $Pb(ClO_4)_2$ solution [261].

Total sulphur in fuels and organic substances can be determined after Schöniger-flask combustion [137, 138] and absorption of the gaseous products in sodium nitrite solution (not in peroxide, as this poisons the Pb^{2+}-ISE [430]). Sulphur dioxide is thus oxidized to sulphate, the solution is acidified to pH 1.5–2.5, excess of NO_2^- and CO_2 is expelled by boiling, the pH is adjusted to 4.0–6.5 (dil. NaOH solution), alcohol is added and the sulphate is titrated with $0.01M$ $Pb(ClO_4)_2$. Sulphur is determined analogously in petroleum [431] and coal [432]. Sulphate can be determined in steel, petroleum products and aerosols after reduction to sulphide [432a].

Organic sulphates that are soluble in 50–70 % aqueous alcohol or dioxan can be titrated directly, e.g. in some pharmaceuticals (sulphates of atropine, ephedrine, papaverine, and other drugs) [433].

4.6.3 Determination of Sulphur Dioxide

Sulphur dioxide, as well as sulphite and bisulphite, are determined by the gas-sensing electrode after acidification of the test solution. The electrode sensitivity depends on the reaction

$$SO_2 + H_2O \;\rightleftharpoons\; H^+ + HSO_3^- \qquad (4.14)$$

taking place in the internal electrolyte in contact with a pH-sensitive electrode, i.e. on the internal electrolyte concentration (Fig. 2.5). The value of the H_2SO_3 dissociation constant ($pK_{a1} = 1.81$ at 18 °C [221]) indicates that a pH value of about zero is required for complete conversion of SO_3^{2-} and HSO_3^- into SO_2. To avoid addition of a large amount of acid, it is recommended to adjust the solution pH to *ca.* 1.8 to 1.9 with a concentrated sulphate-bisulphate buffer [351]. The SO_2 concentration is then about half that at pH = 0, but there are no substantial differences in the sensitivity. The detection limit has been reported to be $3 \times 10^{-6}M$ (Orion 95-64 [83]) or $5 \times 10^{-6}M$ (EIL Sulfur Dioxide Probe, Model 8010-2 [434]).

Hydrochloric acid ($> 1M$), hydrofluoric acid, and gaseous Cl_2 and NO_2 interfere. Because of the volatility and easy oxidation of SO_2, the measurements are performed in at least partially closed vessels or in flow-through systems. Standards are prepared from sodium sulphite and the electrode is stored in $0.05M$ HSO_3^- solution.

The method has been applied to the determination of SO_2 in fuels, e.g. petroleum [435]. The sample is combusted and the products are absorbed in an $HgCl_4^{2-}$-EDTA solution. SO_2 displaces the Cl in the complex and the Cl^- released is measured. Sulphamic acid is added to remove NO_2^-; the measurement is then performed in Na_2SO_4 + H_2SO_4 buffer at pH 1.2. Sulphur dioxide in the atmosphere [435, 436] and smoke and exhaust gases is determined analogously. The SO_2 concentration can also be determined indirectly in gaseous mixtures, with an I^-- or Pb^{2+}-ISE. With the first [212] (see also [436a]) the gas is passed through a solution of iodine, the remaining iodine is removed by extraction into carbon tetrachloride and the iodide formed is determined. In the second procedure [437], SO_2 is oxidized by hydrogen peroxide and the sulphate formed is titrated with Pb^{2+}.

Sulphur dioxide has also been determined in dried foodstuffs [83, 438] and in wine [83, 438, 439], in a medium containing 20 ml of conc. HCl and 10 ml of glycerine per litre.

4.7 DETERMINATION OF NITROGEN COMPOUNDS

4.7.1 Determination of Ammonia

Determination of ammonia and of total nitrogen after its reduction to ammonia is important in biochemistry and in the analysis of waters and the atmosphere. Ammonia-sensitive electrodes further permit the determination of many non-ionic organic compounds, e.g. urea and amino-acids.

Three types of ISE are available for the determination of ammonium ions: glass electrodes, ISE's based on nonactin and monactin, and the gas-sensing NH_3 electrode. Glass electrodes [440] (the best-known are Beckman 39137, EIL 1057-2 and Corning 47622) do not differentiate between NH_4^+ and K^+ and generally exhibit poor selectivity (e.g. with respect to H^+ and Na^+). Therefore, they are used only under extreme conditions (e.g. at higher temperatures and pressures) [441]. Commercial NH_4^+-LISE's contain 72 % nonactin and 28 % monactin in tris(2-ethyl-hexyl)phosphate [442] (Philips IS 560) and were originally designed for the determination of potassium ions [443]. This electrode can be employed in a concentration range of 10^{-1}–$10^{-5}M$ NH_4^+ ($k_{NH_4^+, K^+}^{Pot} = 0.1$ and $k_{NH_4^+, Na^+}^{Pot} = 2 \times 10^{-3}$ [442]) and is applied e.g. in the automated determination of ammonia in boiler-feed waters [444] or in the detection of organic NH_4^+-derivatives [445] (glass electrodes are also sensitive to these derivatives [446]).

The gas-sensing NH_3-ISE is one of the most selective electrodes (Orion 95-10 or EIL Model 8002-2). It is based on monitoring changes in the pH of an internal electrolyte ($0.01M$ [351] or $10^{-3}M$ NH_4Cl [447], see also [448]) due to the reaction

$$NH_3 + H_2O \rightleftharpoons NH_4^+ + OH^- \qquad (4.15)$$

Either the membrane is in contact with the test solution [351] or the air-gap electrode [447, 449] is used (see Fig. 2.4). The most recent type employs monitoring of Ag^+, Cu^{2+} or Hg^{2+} instead of the pH in the internal solution [450]. This electrode has a somewhat higher detection limit (above $10^{-4}M$), but its response is faster and the slope may be higher than that of the commercial electrodes ($\sim 100\,\text{mV/log}\,[\text{NH}_3]$). The pH of the test solution is adjusted to 11.2 and the temperature and the ionic strength (max. 0.2) are maintained constant. The detection limit is *ca.* $10^{-6}M$ NH_3 (17 µg/l.) [451, 452].

The high selectivity for the determination with the gas-sensing ISE is given by the impermeability of the membrane to most ions and the fact that most other gases, such as CO_2, H_2S and SO_2, exist in solution in the ionic form at pH $\geqq$ 11.2 (CO_3^{2-}, HS^-, SO_3^{2-}). Only certain amines, e.g. cyclohexylamine and octyldecylamine [453], and some substances wetting the membrane (when the air-gap electrode is not used) interfere. Mercuric ions can be removed by adding sodium iodide.

After replacement of the membrane, it is recommended to test the electrode in pH-buffers. We repeatedly observed relatively short lifetimes (maximum 6 months) with the Orion 95-10 electrode during measurements under the conditions recommended by the manufacturer [454]. Standards are prepared from NH_4Cl; the electrode is stored in $0.1M$ NH_4Cl [454].

The ammonia electrode is primarily used for determinations in various waters, which are conserved by addition of 2 ml of conc. HCl after sampling. The measurement is performed on 50–100 ml of the sample, after addition of 0.5–1 ml of $10M$ NaOH. NH_3 concentrations higher than *ca.* 0.02 mg/l. can be determined. Chlorinated waters are freed from chlorine by adding $Na_2S_2O_3$. Determinations in waste [447, 451, 455], potable [456], mineral [451] and industrial [457] waters have been described. The procedure has further been applied to the determination in boilers [441, 444], boiler-feed waters [453], fish-tanks [458] and aquaria [455]. Soil extracts are analysed analogously [459, 460]. The total nitrogen in soils can be determined after Kjeldahl decomposition of the sample [459]. In the determination of NH_3 in the atmosphere, the air is passed through an acidic absorption solution (e.g. $0.1M$ HCl) or the particulate matter is deposited on a glass fibre filter. The filter is extracted with water and made alkaline (NaOH) before the determination [461]. For NH_3 in air, it is possible to determine 2600–0.03 µg/m^3.

Ammonia or the total nitrogen is frequently determined in urine [462, 463], blood [464] and blood serum [465]. Animal faeces are extracted with water or $0.1M$ HCl, the mixture is centrifuged and the solution is made alkaline [466].

Ammonia has also been determined in wine and cider [467], beer [468] and grain, starch, fodders, cheeses and meat products. The latter materials are kjeldahled, diluted and made alkaline [469]. Ammonia can be determined by direct potentiometry in plant extracts and fodder solution [460a]; crude protein in plant material [470] and primary carbamates [471] can also be determined.

4.7.2 Determination of Urea, Amino-Acids and Other Nitrogen-Containing Substances by Using Enzymatic Reactions

The determination of ammonia and urea is important not only in waste products, but chiefly in blood. Urea is subjected to enzymatic hydrolysis,

$$CO(NH_2)_2 + 2\,H_2O + H^+ \xrightarrow{\text{urease}} 2\,NH_4^+ + HCO_3^- \quad (4.16)$$

and the concentration of NH_4^+ is monitored. The enzyme can be added directly to the sample [472, 473] and the NH_4^+ concentration measured after a certain incubation period (e.g. 10 min [473]). More frequently [474 – 476], enzyme ISE's [476] are used, employing an enzymatic reactor [476a] or with urease immobilized on the surface of a glass [474, 475], nonactin [477] or gas-sensing [478] ammonia-ISE. With the last electrode, the difficulties caused by the interference of the alkali metal ions are avoided, but the porous membrane is often clogged during measurement in biological fluids; therefore, the use of the air-gap electrode [439, 476, 479] with a linear response to urea within a concentration range of $5 \times 10^{-2} - 1 \times 10^{-4} M$ [479] seems to be most suitable. The enzyme is fixed on the vessel bottom [476] or on the magnetic stirring bar [479]; tris-buffer of pH 8.5 is used [0.5M tris-(hydroxymethyl)aminomethane + HCl]. Urea can also be determined with the carbon dioxide electrode, indicating the HCO_3^- formed during reaction (4.16) [480]. The measurement can be automated [473]. It is possible to determine both NH_3 (NH_4^+) and urea in animal waste products [462] by measuring first NH_3 in alkaline solution and incubating the second aliquot with urease in a tris-buffer of pH 7, making the solution alkaline and measuring the NH_3 again; the urea is determined from the difference. The determination of other nitrogen-containing substances in animal excrements, e.g. creatinine after hydrolysis with creatinase [481] or arginine [480] is analogous.

Various amino-acids [482–485c], e.g. lysine, tyrosine, leucine, phenyl-alanine, can also be hydrolysed enzymatically and determined by monitor-ing the ions formed, by using ammonia or carbon dioxide ISE's. L-argi-nine and L-lysine have been determined by using a CO_3^{2-}-ISE [485d]. Amino-acid-responsive LISE's based on quarternary amino-acid salts have also been described [486]; these electrodes do not require enzymatic hydrolysis. Argentometric titration with the Ag_2S-ISE was employed for the determination of p-urazine in $1M$ NaOH [487].

4.7.3 Determination of Nitrite and Nitrogen Oxides

The NO_2/NO_2^- gas-sensing electrode is analogous to the NH_3-electrode, as are the measuring techniques: $0.02M$ $NaNO_2$ is employed as the internal electrolyte; the detection limit is *ca.* $5 \times 10^{-7}M$ NO_2 [351]. (The NO_2-ISE made from a chalcogenide glass [488] is based on a different principle.) The test solution has a pH of 1.1–1.5 and the ionic strength is maintained constant by adding a buffer containing 190 g of Na_2SO_4 and 53 ml of 97 % H_2SO_4 in 1000 ml of water [489]. Weak volatile acids such as acetic, formic, hydrofluoric, lactic or pyruvic acids and CO_2 when present in 30-fold amount relative to NO_2, interfere. Substances reacting with HNO_2 (Cl^-, Br^-, I^-, Sn^{2+} and Fe^{2+}, amines and substances that can undergo diazotization) also interfere.

Most frequently, NO_3^--ISE's are employed for the determination of nitrogen oxides and nitrite, after their oxidation to nitrate, mostly with $KMnO_4$ or H_2O_2 present in the absorption solutions [490–492]. Nitrogen oxides can also be oxidized with ozone [493] or with PbO heated to 190 °C in a borosilicate tube: $Pb(NO_3)_2$ is formed, which is dissolved in a phosphate buffer of pH 11.5 [494]. These procedures are usually applied to the analysis of aerosols, the atmosphere and various gaseous emissions and have also been automated [514]. The NO_2/NO_2^--ISE was used for the determination of nitrite in soil extracts [495] and of nitrogen oxides in the atmosphere after their absorption in a buffer containing 3.7 g of NaH_2PO_4, 3.3 g of Na_2HPO_4, 10.86 g of $HgCl_2$, 5.9 g of KCl and 2.9 g of $K_2Cr_2O_7$ in 1 litre (pH 6.2); the buffer also removes the SO_2 interference [489]. Nitrite can be determined in solutions used for pickling meat and in foodstuffs in concentrations higher than 1 ppm. Because formic and acetic acids in these materials may interfere, sulphamic acid is added to the solution after the measurement, to destroy NO_2^-, and the measurement is repeated. The nitrite content is then calculated from the difference [489].

4.7.4 Determination of Nitrate

The nitrate content in waters provides important information on their pollution by fertilizers and by decomposition products of organic materials. The determination of nitrate is also important in soils and generally in agriculture.

Nitrate-ISE's resemble the perchlorate-ISE's, as far as the electro-active substances and selectivities are concerned. The most frequently employed electrode is the Orion 92-07, containing an ion-association complex of tris(substituted 1,10-phenanthroline)-nickel(II) with nitrate in an aromatic nitro-compound as solvent (e.g. p-nitrocymene) [17, 496].

$$\left(Ni \left[\begin{array}{c} N \\ N \end{array} \overbrace{\hspace{2cm}}^{} -R \right]_3^{2+} , \quad 2\,NO_3^- \right)$$

Model 92-07 has, however, been replaced by model 93-07, the potential of which is more stable and reproducible.

Another group of NO_3^--ISE's is based on nitrates of substituted ammonium or phosphonium salts with long carbon chains [218, 499, 500, 500a] (e.g. tridecylhexadecylammonium nitrate dissolved in n-octyl-o-nitrophenyl ether is the basis of the Corning 476134 electrode [497, 498]). ISE's containing basic dyes have selectivities similar to the Orion 92-07-ISE, but their impedance is lower [500b, c]. PVC-membrane [498], liquid-state [501] and coated-wire [490] electrodes have been described.

The Nernstian response ranges of PVC-membranes with the two ion-exchanger types are roughly identical (Corning, 10^{-1}–$10^{-4}M$; Orion, 10^{-1}–$2 \times 10^{-4}M$ [498]). According to the manufacturers, the detection

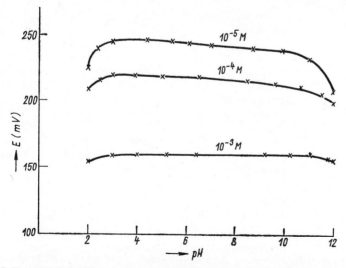

Fig. 4.7 E vs. pH dependence for the Orion 92-07 nitrate electrode in $0.05M$ K_2SO_4. Stirred solutions, 21 °C, potentials were read after 2 min.

limit amounts to $6 \times 10^{-6} M$ NO_3^- for the Orion 93-07 electrode and $10^{-5} M$ for the Corning 476134 electrode. The deviations from the slope reported in the literature are larger than is usual with ISE's and the potential-stability is poorer [498, 501, 502]; therefore, the electrode must be standardized more frequently. The useful pH-range, 2–12 [498] (for the pH-dependence see also [503]) becomes narrower with decreasing NO_3^- concentration (see Fig. 4.7.), so that the lowest activities are best determined at pH 2.8–4.2 [502]. The electrode based on tetradecylammonium nitrate in PVC can be used in very acidic media. (HNO_3 can be determined selectively in the presence of HCl, H_2SO_4 and H_3PO_4 in the range from 10^{-4} to $6.2M$ [499].)

The determination with the NO_3^--ISE is not very selective, as follows from the selectivity coefficients given in Table 4.7. Serious interference

Table 4.7

Selectivity coefficients, $k_{NO_3^-, X}^{Pot}$, for nitrate electrodes

Interfering anion	Electrode				
	Orion 92-07	Corning 476134	Beckman 39618	PVC-membrane	
				Corning 476134	Orion 92-07-02
ClO_4^-	$10^3 a$	$> 10^3$	95.5 44.6b	800c	550c
ClO_3^-	2a 1.14—1.45d 0.89e	—	1.1	1.66c	1.66c
I^-	20a 2.7—22.2d	25	5.6 6b	17c	16c
Br^-	0.9a 0.09—0.24d	0.011	0.28 0.4b	—	—
NO_2^-	0.06a 0.09e	—	0.066	0.066c	0.06c
CN^-	0.02		0.02	—	—
Cl^-	$6 \times 10^{-3} a$ $8 \times 10^{-3} e$	4×10^{-3}	0.02 0.15b	5×10^{-3}	4×10^{-3}
F^-	$5 \times 10^{-4} e$ $9 \times 10^{-4} a$	—	6.6×10^{-3}	8.7×10^{-4}	7×10^{-4}
SO_4^{2-}	$6 \times 10^{-4} a$	10^{-3}	10^{-5}	$< 10^{-5}$	$3 \times 10^{-4} c$

The selectivity coefficients increase with solution dilution [272, 502] and can serve only for approximate assessment of the degree of interference. a ref. [17]; b ref. [288]; c ref. [498]; d ref. [272]; e ref. [503]; for other data see the manufacturer's literature.

is encountered in the presence of ClO_4^- and ClO_3^- (which rarely occur in common materials), and also iodide, fluoroborate and thiocyanate. The electrode is actually more sensitive to all these ions than to nitrate and prolonged exposure to them may cause anion-exchange in the ion-exchanger and poisoning of the electrode; the electrode is then regenerated by soaking in $10^{-3}M$ KNO_3 or by membrane exchange, if the construction permits it. Commonly occurring anions, such as Cl^-, HCO_3^-, Br^-, CN^-, S^{2-}, NO_2^- and anions of organic acids interfere less seriously and their effect can often be removed or at least decreased by relatively simple operations. For example, chloride can be deposited on a cation-exchanger in the Ag^+-form, HCO_3^- removed by acidifying the solution to pH < 4 or on a cation-exchanger in the H^+-form [504]. Chloride and bicarbonate, which normally interfere at nitrate concentrations below 50 µg/l., can then be present in a thousandfold ratio to nitrate in the initial sample. Chloride, cyanide, bromide and iodide can also be precipitated by Ag^+ [505], added e.g. in the form of the buffer, $0.01M$ Ag_2SO_4, $0.01M$ $Al_2(SO_4)_3$, $0.02M$ sulphamic acid and $0.02M$ H_3BO_3 (sulphamic acid suppresses the effect of high nitrite concentrations, while Al^{3+} complexes the anions of organic acids [506]). As this buffer causes an increase in the detection limit by a factor of about 5 and poorer measuring reproducibility when NO_3^- concentrations below 30 mg/l. are measured, it is advantageous to add an Ag_2O suspension instead of Ag_2SO_4 [502]. The NO_3^--ISE can be used for end-point detection in titration of $0.01-0.1M$ nitrate with diphenylthallium(III) sulphate in solutions brought to pH 2-4 with sulphuric acid [507]. Perchlorate, halides and nitrite interfere; nitrite is reduced by hydrazinium sulphate and halides are precipitated by Ag_2SO_4. Samples of waters are conserved with 1 ml of chloroform or 1 ml of 1 % phenylmercuric acetate solution per litre [507a]. Rain waters are acidified with dil. H_2SO_4 or H_3PO_4 to pH 3 and the ionic strength is adjusted with 5 ml of $0.5M$ K_2SO_4 or K_2HPO_4 or with commercial Ion Strength Adjustor solution. Nitrate can then be determined in concentrations above 0.6 mg/l. With natural waters, the buffer is mixed with the sample in a 1 : 20 ratio [505, 506]. Strongly mineralized waters must generally be diluted before the measurement; barium chloride is added to samples with high sulphate concentrations.

Soil samples are extracted with distilled water [506 − 508], $Ca(OH)_2$ solution [509], saturated $CaSO_4$ solution, dilute $CuSO_4$ solution [510], a mixture of dilute $CuSO_4$ and Ag_2SO_4 solutions [508], $0.25M$ K_2SO_4 [511] or $CaCl_2$ [460a, 507]. Solutions of $CuSO_4$ also suppress microbial activity in the test solution (phenylmercuric acetate has the same effect). In analyses of soils with higher contents of clay components which

Ag_2SO_4 solution [83] or a buffer. Plant tissues can be extracted with $0.025M$ $Al_2(SO_4)_3$ and $7.1 \times 10^{-4}M$ NO_3^- at pH 3: the addition of nitrate makes the ISE response more rapid [515]; nitrate can be determined in a concentration range of 1–600 ppm. Chloride and organic acid anions are removed from plant extracts by addition of a cation-exchanger in the Ag^+- and Al^{3+}-forms, respectively [516]. The nitrate electrode is also employed for monitoring the reduction of NO_3^- in microbial cultures [515]. Nitrate nitrogen can be determined in organic materials by using the gas-sensing NH_3/NH_4^+ electrode after Kjeldahl decomposition of the sample and NH_3 distillation [469].

4.8 DETERMINATION OF PHOSPHORUS COMPOUNDS

4.8.1 Determination of Phosphate

Attempts to prepare an ISE selective to phosphate have not been very successful [534–536]. The results obtained with $BiPO_4$ in a PVC-matrix [536a], Ag_3PO_4 on a conductive support [536b, c] and silver phosphates in a mixture with chalcogenides [536d] are more promising, although the theoretical slope (19.72 mV/25 °C) for tervalent ions does not permit a high precision of measurement. Phosphates are mostly determined indirectly by titration with end-point detection by means of F^--, Ag_2S-, AgI- or Pb^{2+}-ISE's. Phosphates in natural waters are precipitated by using a known amount of $La(NO_3)_3$ standard solution and the excess of La^{3+} is titrated with NaF standard solution according to Gran [537]. Similarly, phosphate in concentration higher than 48 µg/l. is precipitated with Ag^+ in an acetate buffer (pH 5.5) and the excess of Ag^+ is determined with Ag_2S- or AgI-ISE's [538].

Organic samples can be combusted according to Schöniger [137, 138] and the phosphate directly titrated in the absorption solution, using the Pb^{2+}-ISE. Ammonium acetate is added before the titration and the pH is adjusted to 8.9 with ammonia. The determination of phosphate is not disturbed by 250 times as much SO_4^{2-} and 100 times as much NO_3^-, but a 100-fold ratio of Cl^- and F^- to PO_4^{3-} causes a positive error. Silicate must not exceed the PO_4^{3-} concentration [539]. It is possible to determine PO_4^{3-} down to 0.01 mg/l.

4.9 DETERMINATION OF CARBON COMPOUNDS

4.9.1 Determination of Carbon Dioxide and Carbonate

There are two ISE types for the determination of inorganic carbon in solution $(CO_2/CO_3^{2-}/HCO_3^-)$, the gas-sensing electrode and the CO_3^{2-}-LISE.

The gas-sensing CO_2 (pCO_2) electrode is the oldest electrode of this type [351, 540−542] and is manufactured in several versions. It is used primarily for estimation of the acid–base condition of blood [543, 544], by measuring pCO_2 and pH values and relating them to a certain standard blood state. The Radiometer ABL-1 instrument measures the blood acid–base balance automatically (pH, pCO_2 and pO_2 electrochemically, haemoglobin photometrically and the barometric pressure).

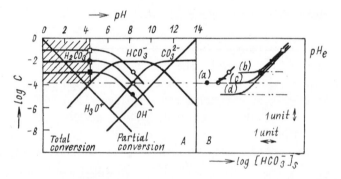

Fig. 4.8 (A) Logarithmic diagram (Hägg diagram) for the carbonate system: pK_{a1} = 6.37, pK_{a2} = 10.25. (B) Calibration curves for a CO_2-electrode with an electrolyte solution with $[HCO_3^-] = 10^{-2}M$, operated by (a) the partial conversion method and (c) the total conversion method. Curves (b) and (d) depict the calibration curves obtained by the total conversion method for electrolyte solution concentrations of 10^{-1} and $10^{-3}M$, respectively. (Reprinted from ref. [545] by permission of Elsevier Publishing Co.).

The CO_2 gas-sensing ISE measures changes in the pH due to diffusion of CO_2 into an HCO_3^- internal solution with a concentration of $10^{-2}M$ [351] or $5 \times 10^{-3}M$ [545]. The electrode exhibits a relatively limited concentration range and the detection limit is ca. $10^{-4}M$ [351, 545]. The determination of lower concentrations is prevented by the presence of CO_2 dissolved in water $(1–3 \times 10^{-5}M)$ [545]. From the dissociation constants of H_2CO_3 (pK_{a1} = 6.37 and pK_{a2} = 10.25 [221]) it follows that the optimum pH for the determination of CO_2 is below 4.4 − see Fig. 4.8 (the pH is adjusted e.g. by $NaHSO_4$).

In addition to inorganic carbon, the total carbon can also be determined, after oxidation of the organic carbon to CO_3^{2-} by persulphate [545]. In view of the high detection limit, samples of less polluted waters ($< 2mM$ total C) must first be preconcentrated by precipitating the carbonate as $PbCO_3$; then it is possible to determine down to $0.1mM$ total carbon.

The CO_3^{2-}-, HCO_3^{-}- and LISE's contain high molecular-weight substituted ammonium salts as electroactive substances [546 – 548] (e.g. 5 % Aliquat 336S in trifluoroacetyl-p-butylbenzene). An electrode of this type was employed for continuous monitoring of the CO_2 level in serum and plasma without interference from Cl^- ions [547, 548a]], in a medium buffered at pH 8.4. A microelectrode has been constructed for CO_2 determination in blood cells [549]. The standardization is carried out with a physiological solution (Na^+ 141 meq/l., HCO_3^- 30 meq/l., Cl^- 116 meq/l., and K^+ 5 meq/l.) at three different CO_2 concentrations and barometric pressure of 20, 40 and 100 mmHg.

4.10 DETERMINATION OF BORON

Boron can be determined after conversion into BF_4^- with concentrated hydrofluoric acid. The ISE described first by Carlson and Paul [550] is most frequently employed; its properties and selectivity are between those of the perchlorate and nitrate-LISE's. LISE's based on Brilliant Green [551] and tetraphenylphosphonium ion [551a] have also been proposed. The electroactive system of the Orion 93-05 electrode is

Measurements with the BF_4^--ISE can be carried out in neutral media (pH 6–7), e.g. in the analysis of borosilicate glass [552], or in HF medium (e.g. 0.28M [550]). Down to $10^{-5}M$ boron can be determined in waters by a procedure involving first deposition of boron on a column of boron-specific ion-exchanger Amberlite XE 243, elution of interfering anions (especially NO_3^-, Br^- and I^-) with dilute ammonia, conversion of the boron into BF_4^- by passage of a small volume of 10% HF solution, removal of excess of HF by washing with water after 15 min, and finally elution of BF_4^- by NaOH. If the eluate is passed through a cation-exchanger in the H^+-form, a mixture of tetrafluoroboric and dilute hydrofluoric acids is obtained which can be measured with the BF_4^--ISE. The method was also applied to the determination of boron in soils, soil extracts and plants [553].

In the determination of boron carbide in aluminium oxide the sample is first fused with Na_2CO_3 + NaF mixture at 950 °C. The melt is extracted with $6M$ HCl and an aliquot of the suspension is measured after neutralization to pH 4.5 with NaOH solution; 0.5–5 % B can be determined [554].

Samples of silicon for the determination of boron [555] are dissolved in $4M$ HF + $2M$ NH_4F and the solution is neutralized to pH 4.8 with ammonia before the measurement. It is possible to determine 0.15–3.45 mg/l. Fluoroborate can also be titrated with tetraphenylarsonium chloride, best at 2 °C, where the potential break is sharpest [293].

4.11 DETERMINATION OF ALKALI METALS

4.11.1 Determination of Lithium

Glass electrodes [556, 557] can be employed or an indirect determination can be carried out by titrating with NH_4F in an anhydrous ethanol medium, using the F^--ISE [558]. A heterogeneous electrode containing vanadium bronzes ($Li_xV_2O_5$; $0.2 < x < 0.6$) [559] and electrodes with electroneutral ion-carriers [560, 560a] have been proposed. Sodium and hydrogen ions interfere in the determination with glass electrodes; therefore, the pH is adjusted to 8–9 with potassium hydroxide or, better, with organic bases. There is no special Li^+-ISE manufactured commercially, but Na^+-glass electrodes can be employed.

4.11.2 Determination of Sodium

4.11.2.1 SODIUM-SELECTIVE ELECTRODES

Potentiometric determination of Na^+ with glass electrodes is relatively old, as the alkaline error of the pH-sensitive electrodes was utilized for this purpose as early as 1931 [561]. More detailed study of Na^+-sensitive glasses [440, 562–566] brought about commercial production of Na^+-ISE's at the end of the fifties. The membranes of present-day electrodes

are made of either NAS_{11-18} or $LAS_{10.4-22.6}$ glass*. The former
exhibits faster response, while the latter has lower selectivity coefficients
with respect to the most important cations. For example, k_{Na^+, K^+}^{Pot} is
10^{-3}–10^{-4} for $LAS_{10.4-22.6}$ and 10^{-2}–5×10^{-3} for NAS_{11-18} and
k_{Na^+, H^+}^{Pot} is 5–10 for $LAS_{10.4-22.6}$ and 30–40 for NAS_{11-18} [567]. The
properties of Na^+-ISE's vary in dependence on conditioning and on
time; it is therefore recommended to soak them for several hours (or
preferably days) in $0.1M$ NaCl before the measurement.

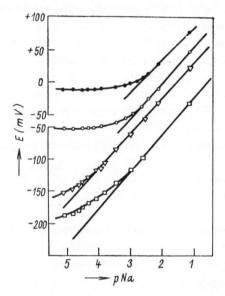

Fig. 4.9 Influence of potassium ion on the potential response of sodium ion-selective
electrodes in $0.1M$ potassium chloride. 25 °C. ● — Beckman 39278 electrodes,
□ — EIL GEA 33C electrodes, ○ — Orion 94-11 electrodes, △ — Radiometer G502 Na
electrodes. (Reprinted from ref. [567] by permission of Elsevier Publishing Co.).

Sodium can be determined only in media with a suitable pH value
(pH > pNa + 3) [568] adjusted with organic bases, e.g. ethanolamine,
triethanolamine, or ammonia, in combination with HCl. Large organic
cations do not interfere. Various commercial electrodes have different
selectivities with respect to H^+; e.g. $10^{-4}M$ Na^+ can be measured at
pH > 6.5 (G 502 Na, Radiometer), pH > 7.0 (GEA 33C, EIL) or
pH > 8 (94-11, Orion) [567]. Silver interferes seriously (k_{Na^+, Ag^+}^{Pot}
= 70–400 for various commercial electrodes [567]); even the Ag^+

---

* The composition of ion-selective glasses is usually given in mole%; i.e. NAS_{11-18}
denotes the composition, 11 mole % Na_2O, 18 mole % Al_2O_3 and 71 mole % SiO_2.

present in solution in equilibrium with an AgCl precipitate prevents the determination of lower Na^+ activities. Therefore, double-junction Ag/AgCl electrodes must be used or a different reference electrode. Lithium ions also interfere, whereas the ISE is relatively Na^+-selective with respect to NH^+ and K^+ (see Fig. 4.9).

The detection limit is usually specified as 10^{-5} or $10^{-6}M$, but there have been attempts to measure down to $10^{-8}M$ Na^+ in boiler-feed water at pH 11.6 [569, 570]. The lowest concentrations are best measured in plastic flow-through systems, as under static conditions the solution becomes contaminated by cations from the external reference electrode and from the vessel walls, especially in alkaline solutions. The precision of the potentiometric determination is somewhat better than that of flame photometry, and the procedure is simpler and more readily automated; on the other hand the response is slower (potential stabilization requires $3-5$ min). The response is retarded by exposure to very dilute solutions; therefore, the electrodes are r⁻ insed with water between measurements, but with a $10^{-3}M$ NaCl solution and the sample. The response time also increases with decreasing temperature and some electrodes are not recommended for measurements at temperatures below 20 °C [567]. The selectivity coefficients also strongly depend on the temperature.

Many microelectrodes have been constructed for biological determinations [571]. In addition to the generally used glass electrodes, electrodes based on neutral ion-carriers have also been proposed [572, 572a, b]. Unlike glass electrodes their response time is not affected by proteins. An Na^+-ISE and the injection technique have been used for analysis of serum [572c].

4.11.2.2 DETERMINATION OF SODIUM
IN INORGANIC AND ORGANIC MATERIALS

Sodium is determined especially in surface waters [569, 573–575], boiler-feed waters [569, 570, 576], in sea [577, 578], mineral [573], waste [575, 578] and irrigation waters and soil extracts [108, 579]. Flow-through autoanalysers have been constructed for the determination of very low Na concentrations in waters [570, 574, 575], in which a mixture of air and NH_3 is introduced into the sample or the water is mixed with a buffer containing 2.14 g of NH_4Cl and 10 ml of triethanolamine per litre. In determinations in sea, mineral and waste waters, simulated standards are employed [83]. Soils are extracted with water or $CaCl_2$, and the extract is centrifuged and neutralized with $Ca(OH)_2$ before the measurement. Interference from K^+ can be suppressed by adding a calcium tetraphenylborate solution [108].

The Na^+-ISE can be applied for analyses of plant materials, e.g. grain, maize and sugar beet [580]; the dried samples are extracted with $0.1M$ magnesium acetate and the extract is filtered before the measurement. A similar procedure is adopted in the analysis of certain pharmaceuticals [581] and saccharine [83]. Sodium has also been determined in smoked fish [582] and wine [83].In analyses of foodstuffs the samples or aqueous extracts are buffered with a pH 10.2 ammonia buffer [582]. Sometimes the sample is only diluted with dil. ammonia, e.g. in the analysis of wine.

In clinical analysis, Na^+ is determined in blood [572, 583, 584], serum [585] and urine [83, 277]. Samples are first diluted with water and the pH is adjusted to 8, e.g. with tris-buffer [83]. A sodium microelectrode has been employed for the determination of the Na^+ activity in animal brains, blood and sweat *in vivo* [440, 587, 587a,]. A number of other clinical and biochemical applications are surveyed in [440, 571].

In industry, the Na^+ concentration is determined during the production of sodium carbonate, in the glass industry [83, 588]. The materials are dissolved in water or dil. HCl and the pH is adjusted with ammonia. Paper-mill liquors must usually be diluted with water and ammonium carbonate added [589].

4.11.3 Determination of Potassium

Commercial K^+-ISE's can be classified into three groups: (*a*) glass electrodes; (*b*) neutral ion-carrier electrodes; (*c*) ISE's based on substituted potassium tetraphenylborate.

Glass electrodes useful for the determination of potassium [440, 556, 590] (e.g. Beckman 39137 or Corning 47622) are based on KAS_{20-05} or NAS_{27-04} glasses. These cationic electrodes do not differentiate between univalent cations (e.g. $k^{Pot}_{K^+, Na^+} = 0.33-0.1$) and since their properties also depend on the electrode history, they are unsuitable for the most common application of K^+-ISE's, namely measurement in blood plasma and in other fluids, where univalent cations other than potassium are present in comparable concentrations. Another commercial electrode employs potassium tetra-*p*-chlorophenylborate dissolved e.g. in 3-nitro-*o*-xylene (Corning model 476132 [591, 592]). This electroactive substance was also tested in PVC membranes [593, 594] and applied to the determination of potassium in serum [592]. The widest application is in intra- and extracellular measurements. For a survey see [234a]. The most important selectivity coefficient, $k^{Pot}_{K^+, Na^+} = 1.2 \times 10^{-2}$, is higher than that for valinomycin electrodes. A similar selectivity is also exhibited by a heterogeneous electrode containing $K_2ZnFe(CN)_6$ ($k^{Pot}_{K^+, Na^+} = 2.5 \times 10^{-2}$ [595, 595a]).

Of course, ISE's based on neutral ion-carriers, especially valinomycin, are most popular, as their selectivity is the highest. These ISE's appeared as a by-product of biological research on ion transport in living organisms [23, 596]. Neutral ion-carriers have, or form on complexation with

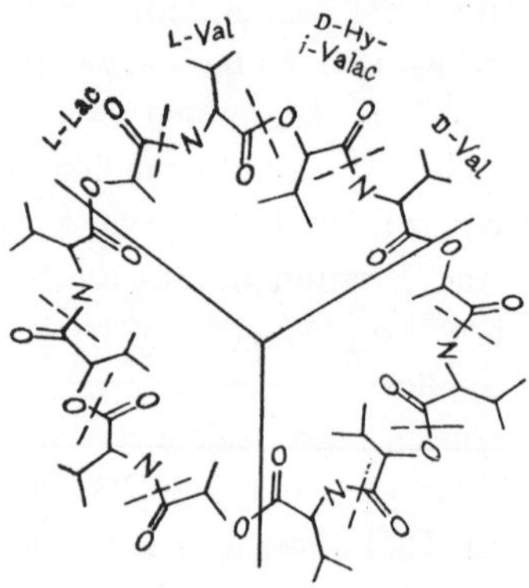

Fig. 4.10 Structure of valinomycin. (Reprinted from ref. [23] by permission of Cambridge University Press).

a cation, a structure with polar bonds oriented into the centre of the complex, while the outer envelope is non-polar and ensures the lipophilicity of the whole aggregate. The stability of complexes with various cations substantially depends on the ligand structure, dimensions of the cations complexed and the minimum dimensions of the cavity inside the ligand; consequently, the complexation is highly selective.

The first substance of this group proposed for a K^+-ISE was the macrotetrolide nonactin [443], which is still used for NH_4^+-electrodes [442]. Another group of compounds useful for preparation of K^+-ISE's are the cyclic polyethers synthesized by Pedersen [597]. The electrode which has so far been most successful contains dimethyldibenzo-30-crown-10, which, built into a PVC matrix, has $k^{Pot}_{K^+, Na^+} = 3-4 \times 10^{-3}$ [598, 599, 599a]. This selectivity suffices for analyses of blood serum and urine [600]; however, compared with similar valinomycin electrodes the stability of the "crown" electrode is poorer and the detection limit higher [601]. A microelectrode has also been described [601a].

Most commercial electrodes contain valinomycin, which is a cyclic depsipeptide, i.e. composed of alternately ordered α-amino-acids and α-hydroxy-acids (see Fig. 4.10). For complexation with K^+ see [23, 602]. The original electrode was an LISE with valinomycin dissolved in diphenyl ether (e.g. a $0.009M$ solution [603], and held in a porous membrane [603–605] (Philips IS 560-K and Orion 92-19). These electrodes had short lifetimes [606] and faults were encountered after cooling to the freezing point [607, 608]; therefore, "solid" membranes were prepared, first containing lipids [609, 610] (see Chapter 2) and then valinomycin in high molecular-weight matrices, mostly PVC [599,

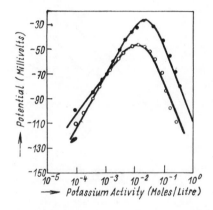

Fig. 4.11 Response of potassium-ISE (valinomycin in PVC). Tributyl phosphate + diphenyl ether plasticizer. Curves denoted as Rb^+, Cs^+, NH_4^+ and Na^+ were obtained in solutions with a constant K^+ concentration of $10^{-4}M$. (Reprinted from ref. [611] by permission of Elsevier Publishing Co.).

608, 611–614]. The advantage of the latter electrodes is a longer life with identical or somewhat higher selectivity which, however, depends on the plasticizer used [608, 615]. In the absence of a plasticizer, Nernstian response is not attained with PVC electrodes [611]. Potassium electrodes without internal reference solutions have also been described [599, 608, 615], as well as the so-called K^+-ISFET (ion-sensitive field-effect transistor) electrode [616] and a microelectrode with a tip diameter of ca. 2 μm [616a]. The response slope exhibited by coated-wire electrodes depends on the thickness of the polymeric layer [599].

Valinomycin electrodes are preconditioned in $10^{-3}M$ KCl for several hours in order to hasten the response. No reference electrodes which contain K^+ in the internal solution may be used; double-junction electrodes with a concentrated NH_4NO_3 solution in a liquid bridge are frequently used. However, the measurement of the lowest K^+ activities can be affected by interference from NH_4^+ ($k_{K^+, NH_4^+}^{Pot} \cong 1-2 \times 10^{-2}$). Solutions of $NaNO_3$ or NaCl may be more suitable in such cases, although a larger liquid-junction potential is involved.

The detection limit with valinomycin electrodes lies between 10^{-5} and $10^{-6} M$ K^+. The slope of the E vs. a_{K^+} dependence may fluctuate considerably, depending on the kind of anion present [598, 603, 605, 606].

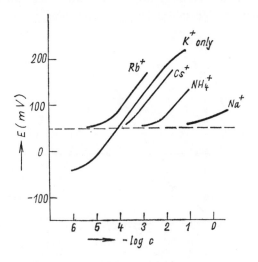

Fig. 4.12 Response of a potassium-ISE in potassium thiocyanate. (Reprinted from ref. [606] by permission of the American Chemical Society).

Lyophilic anions, such as picrate, tetraphenylborate, SCN^-, etc., may penetrate into the membrane, slow down the response and cause anion interference. In concentrated solutions of some of these anions, the response to K^+ may change into an anion response [606] (see Fig. 4.11). This interference, which has not been observed with silicone rubber electrodes [613], can be substantially reduced by adding tetraphenylborate to the PVC-membrane [617].

The selectivity data for valinomycin electrodes (Table 4.9) indicate that, neglecting Rb^+ and Cs^+ which rarely occur in higher concentrations, NH_4^+ is the main interferent if present in more than a tenfold ratio to K^+ (see Fig. 4.12). Interference from surfactants has also been observed [618].

Valinomycin electrodes can be used in buffered and unbuffered solutions, best at pH 4–9. The recommended buffer contains triethanol-

Table 4.9

The selectivity of potassium-selective electrodes, $k_{K^+,x}^{Pot}$

	Beckman 39622	Corning-EEL 476132	Crytur 17-19	Philips IS 560-K	Radiometer valinomcin F 2312K	LISE valinomcin in diphenyl ether	PVC + Valinomycin + plasticizer: TBP + DBP	DOA	di-n-decyl phthalate	PVC + dimethyl-dibenzo-30 crown-10 + DPP
Na+	5×10^{-5}	1.2×10^{-2}	1.1–3.4×10^{-4}	2.6×10^{-4} 9.5×10^{-5} (b)	7×10^{-5}	$\sim10^{-4}$	2–10^{-4}	6×10^{-5}	10^{-3}	3.9–7.2×10^{-3}
NH4+	1.9×10^{-2}	2.3×10^{-2}	1.1–2.2×10^{-2}	1.2×10^{-2} 1.6×10^{-2} (b)	1.9×10^{-2}	2×10^{-2}	1.2–2×10^{-2}	1.3×10^{-2}	1.3×10^{-2}	7.0×10^{-2}*
Li+	3×10^{-4}*	—	7.6×10^{-5}*	2.1×10^{-4} 2.2×10^{-4} (b)	4×10^{-5}	7×10^{-5}	10^{-4}	2.4×10^{-4}	10^{-3}	5.1×10^{-3}*
Rb+	2.2*	10	2.1*	1.9, 2.1 (b)	2.8	2.5	3	2.1	2.5	9.2×10^{-1}*
Cs+	0.5*	20	9.2×10^{-2}*	0.38 0.47 (b)	0.5	—	0.26–0.5	0.47	0.44	2.5×10^{-1}*
H+	2×10^{-4}	—	3.9×10^{-4}	5.5×10^{-5}	—	—	—	—	10^{-3}	—
Ca2+	3×10^{-5}	—	—	4.2×10^{-5} (b)	—	—	—	4.9×10^{-5}	10^{-3}	3×10^{-4}*
Notes	ref. 610	manuf. lit. probably subst. tetraphenyl borate	ref. 601 and 627	manuf. lit. (b) ref. 608	manuf. lit.	ref. 607	ref. 611	ref. 608 "Selectrode"	ref. 615 coated wire electrode	ref. 598

Note: The values marked with an asterisk were measured by the separate-solution method; the manufacturer's literature does not specify the method for the determination of $k_{K^+,x}^{Pot}$; values obtained by the mixed-solution method depend on the level of X and the K/X ratio and are only a guide. Orion Research give the following concentrations causing a 10 % error in the measurement of $10^{-3} M$ K+ with the 93-19 electrode;, $5\times10^{-1}M$ Na+, $3.3\times10^{-3}M$ NH4+, $1M$ Li+, $1\times10^{-4}M$ Cs+ and $1\times10^{-2}M$ H3O+.

amine hydrochloride or acetate and ethanolamine, so that the overall amine concentration is $0.1M$ [613].

Potassium electrodes are most frequently applied to analyses of blood serum [600, 620, 626] (pH 7.3–7.4, $0.145M$ Na$^+$, $0.0042M$ K$^+$, $1.25 \times 10^{-3} M$ Ca^{2+} and $8 \times 10^{-4} M$ Mg^{2+}, $0.103M$ Cl$^-$, $0.027M$ HCO$_3^-$ and small amounts of sulphate and phosphate [619]) and urine [586, 600]. The selectivity is sufficient, but problems may be encountered with the effect of high molecular-weight substances. The samples are usually diluted with water (1:10 for serum and 1:20 for urine) [600]; the sample can be deproteinated with trichloroacetic acid and treatment of the supernatant liquid with Li$_2$CO$_3$, or the proteins can be extracted into chloroform. Automated instruments for these determinations are commercially available (e.g. Stat/Ion of Technicon Instrument Corp. or Orion Biomedical Model SS-30). Various microelectrodes (e.g. [620]) and reference standards [621] have been proposed. Standard pK$^+$ = 2.504 contains 0.1458 mole of NaCl + 0.0042 mole of KCl per kg, i.e. $0.1450M$ NaCl + $0.0042M$ KCl. A survey of clinical applications is given in [571].

The valinomycin electrode can serve for end-point detection in the titration of K$^+$ with sodium tetraphenylborate at pH 5–12; Na$^+$, Li$^+$ and Ca^{2+} can be present in 25-, 50- and 200-fold ratio to K$^+$, respectively [622]. However, NH$_4^+$, Ag$^+$, Rb$^+$ and Cs$^+$ interfere. In the determination of potassium in sea-water, a multiple-addition method with a standard KCl solution gives better results and has been adapted for a computer-controlled titrator [623]. Potassium has been determined in fresh waters with a flow-through analyser [575], in aqueous plant extracts [624] and in biological materials [624, 625]. Plant materials are digested with dilute mineral acids or are boiled in water and the extract is filtered (and its pH adjusted if necessary) before the measurement. Soil extracts are treated analogously [625].

4.12 DETERMINATION OF MAGNESIUM AND OF WATER HARDNESS

There is no commercial electrode designed specifically for magnesium ions; however, the so-called bivalent cation electrode based on phosphoric acid esters dissolved in decanol [17] exhibits similar selectivities for Ca^{2+} and Mg^{2+} and can be employed for the determination of water hardness, magnesium or other bivalent cations (e.g. Be^{2+} or Zn^{2+}) after conversion of the ion-exchanger into the appropriate form.

The selectivity of the bivalent cation ISE, Orion 92-32 (now replaced by the 93-32 type) for Ca^{2+} with respect to univalent cations was poorer than that of the Orion Ca^{2+}-ISE ($k_{Ca^{2+},Na^+}^{Pot} = 0.04$–$0.07$, $k_{Ca^{2+},K^+}^{Pot} = 0.03$–$0.055$ and $k_{Ca^{2+},Li^+}^{Pot} = 0.15$–$0.30$ [628]); the selectivity coefficients were also more dependent on the principal ion level and were higher during the first few days after the LISE preparation [628]. Other bivalent cations, e.g. Pb^{2+}, Zn^{2+}, Be^{2+}, Ni^{2+} and Fe^{2+}, may seriously interfere in the determination of magnesium and water hardness; fortunately, their concentrations in the samples are usually substantially lower than those of the alkaline earth ions. The E vs. pH plots for the Orion 92-32 electrode do not exhibit the dips characteristic of the Ca^{2+}-ISE and the interference from perchlorate and iodide is less serious [629]. The detection limit of the newer electrode, Orion 93-32, is $6 \times 10^{-6} M$ [83].

4.12.1 Determination of Magnesium

The bivalent cation electrode has been employed for the determination of magnesium (or Ca + Mg) in fresh [630−633] and sea [633−635] waters and in soils and for the determination of water hardness. In the titration of Mg^{2+} with DCTA, calcium is masked with EGTA [636], or the first sample aliquot is brought to pH 9.0–9.7 with ammonia or with a glycine buffer, the sum, $Mg^{2+} + Ca^{2+}$, is titrated with EDTA or DCTA and a second aliquot is made alkaline (pH 11–12) with NaOH, thus precipitating Mg^{2+}, and calcium is titrated, using the Ca^{2+}-ISE [637, 638]. Magnesium can also be titrated with triethylenetetramine-hexa-acetic acid (TTHA) in the presence of other cations, e.g. Zn^{2+} or Pb^{2+}. Pairs of cations can be titrated successively when the pH is suitably adjusted [639].

Magnesium can be determined in waters by direct potentiometry, utilizing the large difference in the stability constants of the EGTA complexes of Ca^{2+} and Mg^{2+} ($\log K_{(CaL)} = 11.0$, $\log K_{(MgL)} = 5.2$), so the determination can be carried out in the presence of Ca^{2+} and other cations forming sufficiently stable complexes with EGTA (Ba^{2+}, Fe^{2+}, Fe^{3+}, Cu^{2+}, Zn^{2+}, Pb^{2+}, Ni^{2+}, Co^{2+}, Cd^{2+}, Mn^{2+} and Sn^{2+}), provided that their concentration does not exceed $10^{-3} M$ in presence of $10^{-4} M$ Mg^{2+}. The solution pH is adjusted to 7.0 with diethanolamine or HCl and $10^{-3} M$ EGTA is added [631]; down to $5 \times 10^{-6} M$ Mg^{2+} can be determined. Sea-water must be diluted hundredfold to decrease the NaCl concentration. Soils are extracted with $1 M$ ammonium acetate, the extract is evaporated to dryness, the organic matter combusted, the residue dissolved in $0.5M$ HCl, the solution neutralized with dil. KOH solution to pH 7, 1,2-bis(2-aminoethoxy)ethane-N,N,N,N-tetra–acetic acid added and magnesium measured [640]. This procedure is also applicable to waste waters [635]. Polyamides interfere in the determination of Mg^{2+} at the physiological level [635a].

4.12.2 Determination of Water Hardness

The total water hardness is the total concentration of bivalent metal ions (virtually $Ca^{2+} + Mg^{2+}$), expressed in ppm $CaCO_3$. Carbonate (or temporary) hardness, which can be removed by boiling (calcium and magnesium carbonates are precipitated) and non-carbonate (or permanent) hardness (the difference between the total and carbonate hardness) can be differentiated. The total water hardness can be determined by measuring both the sum of $Ca^{2+} + Mg^{2+}$ activities with the bivalent cation ISE and the specific conductance, and employing a nomogram for correction to obtain the water hardness concentration [641].

An automated determination of total water hardness is based on the following procedure [642]: the non-carbonate hardness (i.e. virtually free Ca^{2+} and Mg^{2+}) is determined by direct potentiometry with a bivalent cation ISE, and a calibration curve (measurement A). Then a known addition of a standard calcium solution is made (measurement B) and the sample is diluted with water 1:1 (measurement C). The total water hardness is then calculated from measurements A, B and C (known-addition known-dilution method − see Chapter 3) by using a nomogram. The total water hardness can also be determined by EDTA or DCTA titration, monitored with a bivalent cation ISE, in solution adjusted to pH 10 with ammonia [630, 643].

4.13 DETERMINATION OF CALCIUM

According to the number of published papers, the Ca^{2+}-ISE is the third most frequently used electrode. The most important application is the determination of $a_{Ca^{2+}}$ in blood serum and other biological materials, but calcium can also be determined in waters and soil extracts.

4.13.1 Electrodes

Calcium-selective electrodes are a good example of development of ISE's toward greater selectivity. In the oldest publications [644] (see also [645, 645a]), poorly selective and practically useless materials were studied, with limited and poorly reproducible response under specific conditions [644]. Calcium-sensitive glass electrodes are also insufficiently selective [646]. The first analytically useful Ca^{2+}-ISE, described by Ross [647], contained an alkyl ester of phosphoric acid [e.g. $0.1M$ di-(n-decyl)-phosphoric acid] dissolved in a solvent with polar substituents, e.g. di-(n-octyl) phenyl phosphonate.

$$
\begin{array}{ll}
n-C_{10}H_{21}-O \\
n-C_{10}H_{21}-O
\end{array}
P
\begin{array}{l}
O \\
OH
\end{array}
\qquad
\begin{array}{ll}
n-C_8H_{17}-O \\
n-C_8H_{17}-O
\end{array}
P
\begin{array}{l}
O \\

\end{array}
$$

di-(n-decyl)phosphoric acid (DDP)   di-(n-octyl) phenyl phosphonate
                                                                      (DOPP)

LISE's containing these substances in a porous membrane were the basis of the Orion 92-20 electrode. DOPP is not merely a solvent, as substantially different selectivity with respect to other bivalent cations is obtained with the ion-exchanger dissolved in n-decanol — the bivalent cation electrode [17]. The presence of DOPP is also important for the selectivity of the PVC membranes developed later, but its role has not yet been completely elucidated [648, 649].

Despite considerable progress, especially in the selectivity, the original Ca^{2+}-LISE still had many shortcomings, e.g. a lifetime of only dozens of days, and much poorer potential-stability compared with solid-state electrodes, and a certain skill was necessary for work with them. Another step ahead was the preparation of PVC membranes with esters of alkyl-phosphoric acids by Moody et al. [649, 650]. These electrodes have a substantially longer life and a somewhat better selectivity. A mixture of 28.8 % w/w PVC + 71.2 % w/w of dioctyl phenyl phosphonate and monocalcium dihydrogen tetra(didecylphosphate) in a ratio of 10 : 1 was found to be the best for preparation of PVC membranes [649]; the properties of these, however, depend markedly on the ratio of the membrane components.

PVC membranes were also used earlier in Ca^{2+}-LISE's containing 2-thenoyltrifluoroacetone and tributyl phosphate as a plasticizer [645, 651, 652]; other versions of electrodes with this ion-exchanger are also possible [653]. According to Shatkay, these electrodes are comparable with the Orion 92-20 electrode [645] and retain the Ca^{2+}-response up to 100–200-fold ratio of Na^+ and NH_4^+ to Ca^{2+} [652].

As the determination in the presence of excess of univalent cations, especially Na^+ (*ca.* 100-fold ratio in biological samples), suffered from difficulties, electroactive substances with a higher selectivity were sought. First Růžička *et al.* [654, 655] (see also [648]) modified the ion-exchanger by introducing an aryl group and employed the ester of di-n-octylphenyl-phosphoric acid (in PVC with DOPP). The new electrode exhibited substantially lower selectivity coefficients for H_3O^+ and other univalent ions and a lower detection limit.

di-n-octylphenylphosphoric acid

Recently Ammann *et al.* [656, 656a, 657] described improved electrodes based on acyclic, lipophilic ligands (neutral ion-carriers), with a structure which can be schematically depicted as e.g.

According to the authors [656], an electrode with this substance in a PVC membrane, with *o*-nitrophenyl n-octyl ether as a solvent and in the presence of sodium tetraphenylborate, surpasses all previously described Ca^{2+}-ISE's, especially in selectivity toward Mg^{2+}, H_3O^+ and Zn^{2+}.

A number of variants of Ca^{2+}-ISE's containing esters of phosphoric acids have also been described [215, 658 – 661]; very frequently these are electrodes without internal reference solutions [658, 659, 661, 661a]

or are microelectrodes [662, 663]. To convert ionic into electronic conductivity in the Růžička "Selectrode", a paste containing mercury,
calomel, KCl and $CaSO_4$ can be employed [654].

4.13.2 Properties of Calcium-Selective Electrodes

The detection limit is determined by the solubility of the organic phase
in water and is given as 10^{-5}–$10^{-6} M$, although activities of $10^{-8} M$ can
be measured in buffered solutions [654, 656]. The potential of Ca^{2+}-
ISE's exhibits drift, so that the measurement of very small activity changes
is impossible [648] and frequent calibration is necessary. The E vs. pH
dependences for ISE's containing esters of phosphoric acid have the

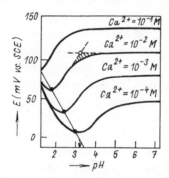

Fig. 4.13 Potential vs. pH plot for a DOPP/PVC electrode at various pCa levels,
showing the influence of pH on the position of the minimum on the curve. (Reprinted
from ref. [654] by permission of Elsevier Publishing Co.).

characteristic shape depicted in Fig. 4.13 [649, 654, 664]; minima appear-
ing in a weakly acidic region at a fixed Ca^{2+} level correspond to the
maximum extraction of the alkaline earths by the phosphoric acids
[654, 664a]. The minima observed with Ca^{2+}-ISE's containing esters of
di-n-dioctyphenylphosphoric acid are located at lower pH values [654].
As a decrease in potential appears at pH 9, apparently because of hydrolysis of Ca^{2+}, the measurements are carried out in approximately neutral
solutions. The k_{Ca^{2+},H^+}^{Pot} value was found to be 4×10^{-5} for the neutral
ion-carrier electrode [656].

The most important selectivity coefficient, k_{Ca^{2+},Na^+}^{Pot}, depends on
many factors, such as membrane loading/formal site concentration and
the ion-mobilities, the activity coefficients and the ion-exchange constants [665] (extremely varied values from 0.42 to 10^{-4} have been given
for the Orion 92-20 electrode [628]). The dependence on the site con-

centration is characteristic of mixed valence systems and is not observed with single-group systems. On prolonged exposure to a concentrated Na^+ solution ($> 10^{-1}M$), gradual substitution of Ca^{2+} by Na^+ occurs in the membrane, which may lead to loss of the response to Ca^{2+}; PVC-membranes can be regenerated by soaking in a $CaCl_2$ solution [654]. These exchange reactions (also with other ions) are apparently also responsible for the poor reproducibility of the measurement in the presence of high Na^+ concentrations. As the Na^+ interference obeys Eq. (1.32), the effect of Na^+ can be reduced by diluting the sample, preserving the $Ca^{2+}:Na^+$ ratio.

According to Hulanicki and Augustowska [628], Na^+, K^+, Br^- and Cs^+ interfere to a low degree with the response of the Orion 92-20 electrode and the selectivity coefficients are independent of the calcium ion activity. However, the selectivity coefficients for the more seriously interfering ions, i.e. Li^+ and NH_4^+, increase with increasing Ca^{2+} activity. The determination with electrodes based on phosphoric acid can also be affected by other bivalent cations, especially Zn^{2+}, Pb^{2+}, Fe^{2+}, Cu^{2+} and Be^{2+}. Higher excesses of Mg^{2+} also interfere by giving rise to temporary responses on the Ca^{2+} background [666]. Anions capable of penetrating into the organic phase, e.g. ClO_4^- or I^-, may also interfere [666]. The response of the Orion 92-20 electrode is further affected by certain organic materials, e.g. phenols, humic acids [667], organic bases and urea [668] and surfactants [669]. The electrode responds to the cyclohexylammonium ion [670].

The interferences found with the Orion 92-20 electrode need not necessarily be observed with other electrodes (see Table 4.10). For example, Zn^{2+} ions do not interfere when the neutral ion-carrier ISE is employed [656]. However, the selectivity of the Orion 93-20 electrode is similar to that of the older type, 92-20.

4.13.3 Determination of Calcium

4.13.3.1 TITRATION

Calcium ions can be titrated with EDTA, DCTA or EGTA, the Ca^{2+}-LISE being used [648, 671 – 674], at pH 9–11 adjusted with an ammonia [633, 673, 674] or glycine [637] buffer. Good results are obtained when $[Ca^{2+}] > 10^{-4}M$ and the Mg/Ca ratio is lower than 100 [632]; magnesium can, however, be masked by adding acetylacetone. EGTA seems to be the best titrant, as it permits selective titration of Ca^{2+} in the

Table 4.10

Selectivity coefficients for calcium selective electrodes, $k^{Pot}_{Ca^{2+}, X}$

	Orion 92-20	Philips IS560-Ca	Radiometer F2112Ca	PVC dioctylphenyl phosphate	PVC neutral ion carrier	PVC Orion ionex 92-20-02
Na^+	3.2×10^{-4} (a) $7 \pm 2 \times 10^{-3}$ (b) 1×10^{-4} (a)*	—	4×10^{-6}	6×10^{-6} (d)	3.2×10^{-4}	6.7×10^{-5}
K^+	$5 \pm 2 \times 10^{-3}$ (b) $0.03 - 0.14$ (b)	—	1.5×10^{-6}	2×10^{-6}*	—	2.2×10^{-5}
Li^+	—	—	6×10^{-5}	6×10^{-5}*	—	
Mg^{2+}	1.4×10^{-2} (a)*	0.032	(c)	2.5×10^{-4}*	4×10^{-5}	$0.220 - 0.036$
Zn^{2+}	3.1 (a)*	$5 - 0.9$	(c)	6×10^{-2}*	1.4×10^{-4}	$0.065 - 0.045$
Cu^{2+}	0.27 (a)*	0.07	(c)	1.6×10^{-4}*	—	—
Sr^{2+}	1.7×10^{-2} (a)*	—	(c)	1.7×10^{-2}*	—	—
Ba^{2+}	1×10^{-2} (a)*	0.02	(c)	2.5×10^{-4}*	—	$0.013 - 0.005$
	(a) ref. 654 (b) ref. 628	manuf. lit.	manuf. lit.	ref. 654	ref. 656	ref. 650

(c) For interference less than 1 mV, the maximum value of the c_X/c_{Ca} is: Mg, 80; Sr, 3; Ba, 500; Cu, 1; Zn, 0.01. For anions the interference is less than 1 mV when the product $c_X \cdot c_{Ca^{2+}}$ is less than: I^-, 10^{-4}; ClO_3^-, 5×10^{-4}; Br^-, 10^{-3}; NO_3^-, 10^{-3}; S^{2-}, 5×10^{-4}. (d) This value was criticized in [656] as somewhat misleading. The maximum interferent concentrations (M) for measurement of $10^{-3} M$ Ca^{2+} with the Orion 93-20 electrode with a 10 % error are given as: Na^+, 0.3; Zn^{2+}, 2×10^{-6}; Pb^{2+}, 5×10^{-6}; Fe^{2+} and Cu^{2+}, 7×10^{-5}; Sr^{2+} and Mg^{2+}, 8×10^{-3}; Ba^{2+}, 3×10^{-2}; Ni^{2+}, 5×10^{-2}.

Note: The selectivity coefficient values depend on the activities of both ions and, to a certain degree, on the measuring method (see refs. 650, 654, 665). The values marked with an asterisk were obtained by the separate-solution method; the method is not specified for results from the manufacturer's literature; other values obtained by a mixed-solution method.

presence of Mg^{2+}. However, the potentiometric titration is generally poorer than the photometric, as the detection limit is higher and the breaks at the end-point are rather small and long drawn-out, especially in the presence of interferents (e.g. Na^+ and Mg^{2+}). This causes difficulties, especially in the analysis of sea-water [632, 648, 671], which can be somewhat alleviated by the use of linear extrapolation [632, 633] or the Gran method [648]. The relative standard deviation is 0.8 % (0.3 % for the photometric titration) [648].

Chelometric titration of Ca^{2+} is better monitored with the Cu^{2+}-[671, 675, 676] or Cd^{2+}-ISE [677], the Cu^{2+}- or Cd^{2+}-EDTA complex being used as a potentiometric indicator. Calcium is titrated either directly in an ammonia buffer at pH 10 or excess of EDTA is back-titrated with a standard Cd^{2+} [83] or Cu^{2+} [677a] solution. Down to $10^{-5}M$ Ca^{2+} can be titrated (Cu^{2+}- or Cd^{2+}-ISE's) even in the presence of high Na^+ concentrations.

4.13.3.2 DIRECT POTENTIOMETRY

The most common application of the Ca^{2+}-ISE is the determination of Ca^{2+} in biological samples, especially blood serum. From the total concentration, 2.477 ± 0.286mM, of calcium in serum, the concentration of Ca^{2+} is 1.136 ± 0.126mM) [678] (see also [679, 680]). For its determination, anaerobic conditions, avoidance of pH fluctuations and addition of anticoagulation agents are necessary. The originally used Orion flow-through 98-20 electrode was replaced in 1975 by the Ion/Stat Biomedical Model SS-20, working automatically at a constant temperature and correcting for the Na^+ concentration. For reference standards see [621].

Buffers are usually unnecessary for the determination of the ionized calcium activity in the presence of protein-bound calcium in serum, as the blood pH (7.3–7.4) is suitable for the measurement. Triethanolamine plus trypsin buffers are sometimes used, as they hasten the electrode response and conserve the sample for 6 hours. However, very pure trypsin must be employed and undissolved particles must be filtered off [682]. A $0.02M$ $NH_4Cl + 0.005M$ NH_3 buffer has been recommended for continuous Ca^{2+} determination in serum [683]. Sodium diethyl-dithiocarbamate is added to the buffer to mask Zn, Pb and other heavy metals. Serum standards are used, e.g. Merck "Titrisol", "Seronorm" or horse serum, and are diluted in a ratio of 1 : 1 with $0.135M$ NaCl to adjust the isotonic salinity. Ca^{2+} concentrations from 0.08 to 400 mg/l. can be determined [683].

The determination of Ca^{2+} in serum is carried out in a blood ultra-filtrate in a dialysis cell. Heparin-treated blood can also be centrifuged, e.g. in a "Vacutainer" tube [681]. If the samples cannot be analysed immediately, they are stored at $-10\,°C$ under a layer of oil. The Orion Ion/Stat does not require sample centrifugation. Dissociation constants of important calcium complexes, e.g. with citrate, EDTA or heparin in heparin-treated blood, can also be determined after addition of a standard amount of $CaCl_2$ with an isotonic salinity of $0.15M$ NaCl.

Table 4.11

Determination of calcium in inorganic and organic materials

Material	Ref.	Material	Ref.
waters	632, 633, 671, 676	urine	695
sea-water	632, 633, 648, 676,	gastric juices	678, 696
	686—688	blood	678, 680, 697—701
soils	640, 685	blood plasma	695, 702—705
clay and bentonite	108, 688	blood serum	621, 676, 679, 681—683,
phosphate rocks	83		699, 703, 704, 706—712
glass	689	cerebrospinal fluid	678, 697, 699,
milk and cream	64, 690		701, 707, 713
beer	684	biological materials	668, 678
flour	83	in cell processes	714, 715
fodders	691	bonding of Ca in an organism	716, 717
sugar solutions	638, 682	measurement of the stability	
saccharine	83	constants of $CaATP^{2-}$ and	
detergents	692, 693	Ca_2ATP (ATP = adenosine	
saliva	694	triphosphate)	718

Calcium in saliva can be determined after dilution of the sample with water or a buffer of pH 6–10. Gastric juices must be neutralized with dilute KOH to pH 6–7. Clinical and biochemical applications of Ca^{2+}-ISE's are surveyed in [678]. In food analysis, Ca^{2+} has been determined in beer after filtration through filter paper in order to equalize the CO_2 concentration with atmospheric CO_2, and neutralization with dil. NaOH solution to pH 5.5–6.0 [684]. Calcium is determined in wine after dilution of the sample with water in 1:9 ratio. Solid materials, e.g. flour or fodders, must first be combusted at $300\,°C$, the ash dissolved in dil. HCl and neutralized with KOH solution to pH 5.5–6.0; for titration with use of the Cd^{2+}-ISE, the sample is brought to pH 10 with ammonia. A similar procedure can be adopted in the analysis of sugar beet, sugar solutions or saccharine [83].

In inorganic analysis, Ca^{2+} is chiefly determined in soil extracts and waters. Soils are extracted with an acetate buffer of pH 8.2 (0.5M sodium acetate) and centrifuged, and the supernatant liquid is diluted with water [685], or 1M ammonium acetate is used as an extractant, the extract evaporated, the organic matter combusted, the residue dissolved in 0.5M HCl, diluted with water and the pH adjusted to 7 [640]. Bentonites and clays are extracted with water, then the extract is diluted with NaCl solution and filtered. White paper-mill liquors are adjusted with ammonia to pH 10, the sulphide present is oxidized with hydrogen peroxide and Ca^{2+} is titrated, the Cd^{2+}-ISE being used [677]. For direct determination of calcium in waters, a buffer of composition 0.4M KNO_3, 0.02M NH_4Cl, 0.02M NH_3, 0.02M acetylacetone and 0.02M sodium iminodiacetate (pH = 9.1), mixed with the sample in a 1:1 ratio, is recommended; 20–800 mg/l. can be determined with an error of *ca.* 4 % rel. It is recommended to deaerate the sample with nitrogen before the analysis [671].

4.14 DETERMINATION OF SILVER

There are many electroactive materials that can be used in Ag^+-ISE's, but the selectivity and sensitivity of the Ag_2S-based electrode are by far the best [17, 375]. Silver-sensitive glass electrodes occasionally find use [440, 721, 722], e.g. in titration of halides [719, 720]. Other Ag^+-ISE types [723–727, 740] have no advantages over these two types. The response slopes of the Ag_2S-based electrode in determinations of Ag^+ are close to the Nernstian value, 59.2 mV/pAg^+ at 25 °C, down to $10^{-7}M$ (0.01 ppm) Ag^+ [322, 344, 728]; deviations can be caused at p$Ag^+ > 5$ by silver adsorption on the electrode and the vessel walls [729], which is probably also the cause of hysteresis effects (the shapes of the E vs. log a_{Ag^+} plots depend on whether the activity is increased or decreased). Adsorption can be decreased by using plastic vessels, preferably made of "Teflon" [729], by rinsing the vessels and electrodes with dil. HNO_3 [344] and by polishing the electrode membrane before replacement of more concentrated test solutions by more dilute ones. Even then the potential-stabilization at p$Ag^+ > 5$ takes several minutes and is very slow at pAg^+ around 7 [322]. In Ag^+-buffered solutions the effect of adsorption is much lower [344, 730] and activities below $10^{-20}M$ can be measured, provided that the total analytical concentration of silver exceeds *ca.* $10^{-6}M$.

The selectivity of the determination is excellent; only mercury cations interfere [344, 731], as can be expected on the basis of the solubility products. The ISE potential is poorly reproducible and in the presence of Hg^{2+} is strongly dependent on stirring of the solution, owing to the formation of an HgS film on the electrode surface; this effect starts at a concentration of 10^{-3}–$10^{-4}M$ Hg^{2+} with $10^{-3}M$ Ag^+ [344]. Chloride at a concentration of $ca.$ $10^{-3}M$ also poisons the membrane surface; the membrane surface ceases to be shiny and the potential exhibits drift. The ISE can be regenerated by brief immersion in $5mM$ NH_3, rinsing with water and pre-conditioning in a $50mM$ solution of sulphide [728].

Silver has been determined in silver-plating baths [261, 359, 732] and in photographic fixers [733]. Preliminary separation of Ag^+ in the form of Ag_2S is recommended, because of the presence of interferents, especially polythionic acids. The precipitate is then dissolved in $1:1$ HCl and an ammonia buffer is added.

Silver can be determined in urine after addition of a drop of dimercaptopropanol (BAL) in toluene to mask Hg^{2+} [734].

The Ag_2S-based ISE is of especial importance in argentometric titrations, e.g. of halides.

4.15 DETERMINATION OF MERCURY

Little attention has so far been paid to development of Hg^{2+}-selective ISE's. The literature mentions heterogeneous HgS and HgSe electrodes in an epoxide matrix [735] with a response slope constant down to $10^{-5}M$ Hg^{2+} and useful down to $10^{-8}M$ Hg^{2+}, silver tetraiodomercurate electrodes with a linear response down to $3 \times 10^{-7}M$ Hg^{2+} [736], the Růžička "Selectrode" with HgS responding to Hg^{2+} [450] coated-wire electrodes sensitive to chloromercurate(II) and iodomercurate(II) [737] and an LISE sensitive to Hg_2^{2+} ion [737a]. The Ag_2S-based electrode gives a non-Nernstian response toward Hg^{2+} [344, 731] that can be utilized in titrations. However, iodide-ISE's are most frequently employed for the determination of Hg^{2+} [329, 504], on the basis of the formation of mercury iodo-complexes with liberation of Ag^+ ions. The E vs. $a_{Hg^{2+}}$ slope is non-Nernstian (47 mV between 10^{-4} and $10^{-5}M$ at pH $=$ 1 [504]) with a detection limit of $ca.$ $10^{-8}M$. At Hg^{2+} concentrations higher than $10^{-6}M$ the electrode surface is blocked, perhaps by formation of HgI_2; the electrode is regenerated by a 10-sec soaking in $10^{-3}M$ NaI and rinsing with water [504].

The iodide-ISE also responds to a certain degree to mercurous ions and to organic compounds of the HgR^+ type, R being an alkyl or aryl group, but not to HgR_2-type compounds [504]. Organic compounds must then be acid-digested and the solution partially neutralized before the measurement. Cyanide, bromide and chloride interfere in the determination of Hg^{2+} by formation of stable complexes; the interference of silver ions ($k_{Hg^{2+},Ag^+}^{Pot} \sim 1$ [504]) can be eliminated by measuring separately at pH > 7, where mercury is precipitated as HgO or is present in the form of hydroxo-complexes. Iodide-sensitive electrodes must not be stored in mercury ion solutions.

Mercuric ions at low concentrations (0.01–1 mg/l.) e.g. in nuclear fuels [738] can be titrated with a potassium iodide solution in dil. HNO_3, the AgI-ISE being used.

Liquid-state mercury electrodes have been developed by Baiulescu and Coşofreţ [738a].

4.16 DETERMINATION OF COPPER

Cupric ion-sensitive ISE's fall into three groups. The first group involves the most frequently used Cu^{2+}-ISE's with pressed or sintered mixed membranes [17, 739]; chalcogenides other than CuS and Ag_2S can also be used [209, 723, 740]. CuS/Ag_2S mixtures also give good results on Růžička's "Selectrode" [741] and in heterogeneous polyethylene membranes [676]. The second group contains electrodes with single-crystal or sintered $Cu_{2-x}Se$ (x is $ca.$ 0.2) [740, 742, 743] or electrodes with copper sulphide [744, 745, 745a], but without silver sulphide. In the third group are other electrodes (e.g. those with chalcogenide glasses obtained by fusion of As_2S_3 with Cu, CuO or CuS [746], an ISE sensitive to $CuCl_4^{2-}$ [746a] or other types [725, 747−749]).

The detection limit in unbuffered solutions is 10^{-7}–$10^{-8}M$ under optimum conditions and down to $10^{-9}M$ Cu^{2+} can be assessed from the rate of the change in the potential of the Orion 94-29 electrode cleansed of adsorbed Cu^{2+} by prior immersion in $0.025M$ H_2SO_4 [750−752]. A flow-through microelectrode was found to yield Nernstian response down to $10^{-8}M$ Cu^{2+}, but the electrode response was slow at very low concentrations [752]. Preconditioning in an EDTA solution is recommended for Růžička's "Selectrodes" [741]; this leads to lower detection limits, but the potential-stability at low activities deteriorates. The potential-stability is also improved by occasional polishing of the membrane [750, 753] or by treatment with silicone oil [753].

The useful potential range is limited by cupric ion hydrolysis in alkaline solutions and by H_3O^+-interference in the acidic region. There is a pH range within which the ISE potential is virtually pH-independent at higher Cu^{2+} activities; however, this range becomes narrower with decreasing Cu^{2+} activity and eventually disappears (Fig. 4.14). The potential becomes more positive with increasing solution acidity, but may decrease on addition of HCl, probably because of complex formation [754]. Halides affect the response of Cu^{2+}-ISE's generally; with Ag_2S-containing electrodes, this effect is perhaps given by the reaction

$$Ag_2S + Cu^{2+} + 2\,Cl^- \;\rightleftharpoons\; 2\,AgCl + CuS \qquad (4.17)$$

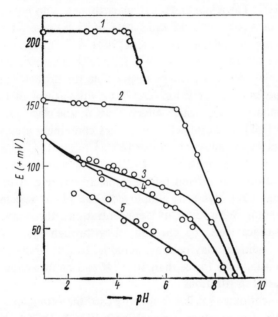

Fig. 4.14 The E vs. pH plot for the Crytur copper-ISE (29-17). (1) $-10^{-1}M\,Cu(NO_3)_2$, (2) $-$ $10^{-3}M\,Cu(NO_3)_2$, (3) $-$ $10^{-5}M\,Cu(NO_3)_2$, (4) $-$ $10^{-6}M\,Cu(NO_3)_2$, (5) $-$ $1M\,NaNO_3$ (deaerated with nitrogen). Ionic strength $I = 1$ ($NaNO_3$). (Reprinted from ref. [742] by permission of Academia).

This reaction would explain the disappearance of the electrode response at $a_{Cu^{2+}} > ca.\ 10^{-3}M$ in $1M$ KCl [751], but the situation is more complicated at high concentrations of Cl^- and low Cu^{2+} activities [746, 751, 754a, b]. This effect also depends on the electrode surface condition [754a, b]. Between 10^{-4} and $10^{-6}M\,Cu^{2+}$ in $1M$ KCl, the Orion 94-29A electrode yields a slope of 35–50 mV/pCu [746, 751, 755]. With heterogeneous Ag_2S/CuS electrodes, Cl^-, Br^- and I^- concentrations must not exceed 10^{-1}, 10^{-4} and $10^{-6}M$, respectively, for measurement of $10^{-3}M$

Cu^{2+} [676]. Chloride and bromide at concentrations higher than $5 \times 10^{-3}M$ and $10^{-3}M$, respectively, interfere with commercial $Cu_{2-x}Se$ electrodes (Radiometer F 1112 Cu and Crytur 17-29) [756). The $CuCl_4^{2-}$-LISE can be used in concentrated chloride solutions [746a]. Other species complexing copper may interfere in determinations with Cu^{2+}-ISE's (e.g. NH_3 or CN^-). Only traces of Ag^+ and Hg^{2+} may be present; ferric and lead ions may also interfere (for $Cu_{2-x}Se$-ISE's, concentrations of Fe^{3+} and Pb^{2+} must not exceed 1/10 and 10 times that of Cu^{2+}, respectively) [754]. The Fe^{3+}-interference can be removed by pH-adjustment to a value above 4, adding NaF to prevent coprecipitation of Cu^{2+} with $Fe(OH)_3$. The alkali and alkaline earth metals do not interfere even if present in a 10^3-10^4-fold ratio to Cu^{2+}; however, the selectivity coefficients are strongly pH-dependent [742]. All commercial Cu^{2+}-ISE's are sensitive to redox systems [742, 753, 757] and to oxygen dissolved in the sample, which is perhaps responsible for drift of the CuS/Ag_2S electrode potential at very low Cu^{2+} activities (through formation of soluble sulphate) [750], which affects the $Cu_{1.8}Se$ electrode potential at pCu > 4.3 [742]. Some mixed membranes containing a copper selenide are reported to be redox-insensitive [757a].

For measurements with Cu^{2+}-ISE's, plastic vessels rinsed with dil. H_2SO_4, and double-junction reference electrodes (e.g. with $2M$ KNO_3 as the bridge electrolyte) are employed. The pH is selected from the Reilley diagram. With low Cu^{2+} concentrations, the cleanliness of the electrode surface must be checked; flow-through systems [752] and anaerobic conditions may be more suitable. In Cu^{2+}-buffered solutions, activities substantially lower than $10^{-10}M$ can be measured [741, 758], which is useful in titrations.

Direct potentiometry is less frequently used than complexometric titration of Cu^{2+} (or of other cations), with EDTA, DCTA, NTA, TEPA (tetraethylenepentamine) and ChelDP [ethylenediaminedi(o-hydroxyphenyl) acetic acid] [675, 741, 759, 760] in an acetate buffer (pH 4.7–5.0) or in an ammonia buffer (pH 10). Concentrations from 10^{-3} to $10^{-5}M$ Cu^{2+} can be titrated. Methanol, acetone, acetonitrile and their mixtures with water can be used as titration media [761]. By direct potentiometry, Cu^{2+} can be determined in natural [757] and sea [741, 746] water. For measurements using addition methods, a buffer of pH 4.8, with composition $0.05M$ sodium acetate, $0.05M$ acetic acid, $0.02M$ sodium fluoride and $0.002M$ formaldehyde, is advantageous when mixed with the sample in a 1:1 ratio [757]. The Gran method permits the determination of Cu concentrations down to 9 μg/l. [83].

Direct potentiometric determination of Cu^{2+} in pyrophosphate [83], acidic metal-plating [261] and etching ferric chloride [83] baths involves dilution of the sample with water or $0.1M$ NaF in $1:99$ ratio and pH adjustment to 6.

One drop of $0.1M$ dimercaptopropanol (BAL) in a mixture of toluene and ethanol is added to urine samples to mask Hg^{2+}, one drop of 10% NaI solution to mask Ag^+ and one drop of acetic acid to adjust the pH to 4–6 and to suppress the Fe^{3+} interference [734].

4.17 DETERMINATION OF LEAD

ISE's sensitive to Pb^{2+} usually contain a PbS/Ag_2S mixture [17, 220, 739, 762–764], but other mixtures have also been proposed, e.g. $PbS/Ag_2S/Cu_2S$ [765] or $PbSe/Ag_2S$ [723, 740, 766, 767]. The presence of silver improves the properties of the Pb^{2+}-ISE [762, 764, 765] and is often recommended. However, PbS heterogeneous electrodes [762, 768], PbS and PbSe single-crystal electrodes [767] and very early on, mineral galena electrodes were also described. The galena electrodes often exhibit lower slopes and fluctuating potentials, but the use of PbSe in a mixture with Ag_2S is recommended for higher stability in acidic solutions and a lower detection limit [766]. Lead telluride membranes exhibit predominantly pH- and redox-sensitivity [767] and resemble elemental tellurium electrodes [770], as can be expected from the highly covalent character of this compound. The best properties are exhibited by sintered electroactive materials (these may also be pressed *in vacuo* and at an elevated temperature) [762, 763, 765, 766]; on the other hand, the presence of $PbSO_4$ leads to great deterioration in the properties [220]. Pb^{2+}-LISE's have also been described [771].

Compared with Ag^+- and Cu^{2+}-ISE's, Pb^{2+}-ISE's show a somewhat poorer reproducibility of potential (± 0.5 mV with the $Ag_2S/PbS/Cu_2S$ electrode [765]). Atmospheric oxygen oxidizes the electrode surface, with loss of sensitivity [762, 764, 772]; therefore, homogeneous membranes must be repolished and the electroactive layer on Růžička's "Selectrode" replaced. The latter electrode should be preconditioned *ca.* 1 day before the measurement, in a $10^{-3}M$ Pb^{2+} solution of pH = 3 and then stored in distilled water [764]. The accessible activity range (ca. 10^{-2}–10^{-6} or $10^{-7}M$) depends on the pH, the purity and the method of preparation of the membrane and other factors. At pH less than 2, lead sulphide slowly dissolves and the chalcogenide ISE potentials are then unstable if pPb > 6. However, in Pb^{2+}-buffered solutions, a response down to pPb $\sim$ 13.5 was observed with the PbS/Ag$_2$S electrode [764, 767], which is in good agreement with the PbS solubility product, $3.4 \sim 10^{-28}$ (18 °C) [221].

Strong interference is encountered from Ag$^+$, Hg^{2+} and Cu^{2+}. Sensitivity to Fe^{3+} was observed with sintered PbS/Ag$_2$S/Cu$_2$S membranes [765] and with PbS/Ag$_2$S heterogeneous ISE's [762]; in the latter case the slope is 59 mV/pFe at pH 1. Higher concentrations of Cd^{2+}, Br$^-$, and, of course, I$^-$ and S^{2-} also interfere. Oxidants, especially peroxides, damage the electrode [430, 762, 764].

Pb^{2+}-ISE's are used almost exclusively for end-point detection in precipitation and complexometric titrations of lead and of a number of anions traditionally titrated with Pb^{2+}(SO$_4^{2-}$, C$_2$O$_4^{2-}$, CrO$_4^{2-}$, WO$_4^{2-}$, P$_2$O$_7^{4-}$, Fe(CN)$_6^{4-}$) [762, 764]. Lead ions can be titrated e.g. with EDTA or NTA in weakly acidic or neutral media [764, 773]. Down to 0.2 mg of Pb^{2+} can be titrated with an error of $\pm 6\%$ in sodium potassium tartrate at pH 5.5–10 [774]. Lead can be titrated with sodium tungstate or molybdate in metal-plating baths [261].

Lead in paints has been determined by direct potentiometry [83], after sample dissolution in HNO$_3$, and neutralization to pH 5–6 with sodium acetate. Attempts have been made to determine Pb^{2+} directly in body fluids in aqueous and non-aqueous media. The best results were obtained with urine [773], where Cu^{2+}, Hg^{2+}, Ag$^+$ and Fe^{3+} were masked with one drop of 1% thiosemicarbazide solution in methanol and Cd^{2+} with one drop of 5% methanolic 1,10-phenanthroline solution [734].

4.18 DETERMINATION OF CADMIUM

Cadmium-sensitive ISE's are similar to Pb^{2+}-ISE's (sintered or pressed mixtures, CdS/Ag$_2$S [739, 776, 777], CdSe/Ag$_2$S [723] or other compounds [777a]). Crystalline CdSe and CdS exhibit sub-Nernstian slopes (17–20 mV/pCd) and their potentials strongly depend on the light intensity [201]. Addition of Cu$_2$S to the CdS/Ag$_2$S mixture makes the response faster and improves the time-stability and mechanical strength [776], which otherwise are a source of difficulties in preparation e.g. of microelectrodes [220]. A CdS/Ag$_2$S mixture can also be used in Růžička's "Selectrode" [778]. The detection limit depends on the pH (pCd $\sim$ 9 at pH 6.7 and pCd $\sim$ 11 at pH ca. 9 in buffered solutions [778]). The value given for the Orion 94-48 electrode in unbuffered Cd^{2+} solutions is $10^{-7} M$, but the actual value is closer to $10^{-6} M$. Prolonged exposure to solutions with pH < 3 may cause a decrease in the sensitivity and changes in the "standard" potential, owing to dissolution of CdS [778]. Silver, mercuric and cupric ions interfere; higher concentrations of Pb^{2+} and Fe^{3+} may damage the membrane, especially in acidic solutions. Among anions, S^{2-}, I^- and higher concentrations of Br^- and Cl^- interfere. For determination in concentrated chloride, a $CdCl_4^{2-}$-LISE must be used [778a]. The applicability of the Cd^{2+}-ISE in direct potentiometry is rather limited and complexometric titrations of Cd^{2+} and other cations (e.g. Ca^{2+}) with NTA, EDTA or 8-hydroxyquinoline in e.g. acetate (pH 4.8), malonate (pH 6.5), ammonia (pH 9.6) or borate (pH 9) buffers are more common. Down to $5 \times 10^{-4} M$ Cd^{2+} can be titrated in ammonia buffer [778]. The 8-hydroxyquinoline titration is best performed in ammonia buffer mixed with dioxan [777] or ethanol [779] (1 : 1). However, the electrode may be passivated by a Cd^{2+}-8-hydroxyquinolinate film, which can be removed only by grinding.

Direct potentiometry can be used in the analysis of cyanide platingbaths [261] and urine [734]. Cu^{2+}, Hg^{2+}, Ag^+ and Fe^{3+} ions are masked in urine by one drop of 1 % thiosemicarbazide solution in methanol and Pb^{2+} and Cu^{2+} by one drop of 5 % thiourea solution in methanol.

In the analysis of natural materials, Cd^{2+} can be extracted selectively into dithizone from a strongly alkaline solution, back-extracted into a mineral acid and determined after pH-adjustment [778].

4.19 DETERMINATION OF VARIOUS INORGANIC MATERIALS

Determinations not discussed in previous sections are summarized in Table 4.12.

Table 4.12

Determination of various inorganic materials

(Abbreviations: AA — analyte addition; AS — analyte subtraction (auxiliary solution is given in brackets); BT — back-titration (in brackets first the reagent, then the back-titrant); DP — direct potentiometry; KA — known addition; KS — known subtraction (auxiliary solution is given in brackets); T — titration (titrant is given in brackets); * — laboratory-made electrode.
Order of entries: (1) method, (2) electrode, (3) medium, (4) interferences, (5) concentration range, (6) other remarks.

Determinand	Procedure	Ref.
Al^{3+}	(1) BT(DCTA, Cu^{2+}), (2) Cu^{2+}- ISE, (4) > 500 times Na^+, NH_4^+, > 20 times Mg^{2+}, (5) 10^{-3}—$10^{-4}M$	780
	(1) AS($10^{-3}M$ F^-), (2) F^-- ISE, (3) acetate buffer, pH 4.6, (5) 10^{-1}—$10^{-4}M$	781
	(1) T(NaF), (2) F^-- ISE, (3) acetate buffer, pH 5 + EtOH	782
	(1: T(NaF), (2) F^- ISE, (3) acetate buffer, pH 3.8—4.5 + $NaClO_4$($I = 0.3$), (4) Fe^{3+}(> Al^{3+}), > $0.01M$ Ca^{2+}, Mg^{2+}, (5) down to $2 \times 10^{-4}M$, (6) Gran plot	537, 783
AsO_3^-	(1) BT(La^{3+}, NaF), (2) F^-- ISE	784
	(1) T(Pb(ClO_4)$_2$), (2) Pb^{2+} - ISE	784
	(1) T($AgNO_3$), (2) Ag_2S - ISE, (3) dioxan — H_2O	784
Au^{3+}	(1) DP, (2) I^-- ISE, (3) NO_3^- or ClO_4^-, pH 1.2—2.3, (4) > $10^{-2}M$ Ag^+, Hg^{2+}, Cl^-, Hg_2^{2+}, > $10^{-3}M$ Bi^{3+}, (5) 10^{-3}—$10^{-8}M$, (6) electrode poisoning at > $10^{-5}M$ Au^{3+}, Pt metals do not interfere	**785**
	(1) DP, (2) $AuCl_4^{2-}$-LISE	785a
Ba^{2+}	(1) DP, (2) neutral ion-carrier ISE, (4) $k_{Ba^{2+}, Ca^{2+}}^{Pot} < 10^{-4}$, (5) 10^{-1}—$10^{-5}M$, (6) *	423
	(1) BT(Na_2SO_4, Pb^{2+}), (2) Pb^{2+} - ISE, (3) mixed	425
	(1) DO, (2) neutral-carrier ISE (PVC)	423a
Be^{2+}	(1) DP, (2) bivalent cation ISE, (6) ion-exchanger converted into Be^{2+}-form	786
Bi^{3+}	(1) AS(I^-), (2) I^- - ISE	787
	(1) T(Na/NH_4pyrrolidinedithiocarbamate), (2) HgS/Hg_2Cl_2 - ISE, (6) *	208
Co^{2+}	(1) DP, (2) $CoSe/Ag_2S$ - ISE, (6) *, slope ~ 25 mV/pCo	723
Co^{3+}	(1) T(Na/NH_4pyrrolidinedithiocarbamate), (2) HgS/Hg_2Cl_2 - ISE, (6) *	208
Cr^{3+}	(1) DP, (2) $NiSe/Ag_2S$ - ISE, (4) Cu^{2+}, Ni^{2+}, Pb^{2+}, Fe^{3+}, Ag^+, Hg^{2+}, I^-, S^{2-}, (5) 10^{-1}—$10^{-6}M$, (6) *, 20 mV/pCr	

continued

Table 4.12 — cont.

Determinand	Procedure	Ref.
CrO_4^{2-}	(1) DP, (2) $(Fe(phen)_3]^{2+}$, $2\ HCrO_4^-$) - LISE, (3) pH 2 — 4, (5) 10^{-2}—$10^{-4}M$, (6) *	291, 788
Cs^+	(1) DP, (2) valinomycin K^+-ISE, (4) K^+, Rb^+	611
	(1) DP, (2) Cs^+-ISE, (6) *	789
	(1) DP, (2) Cs 12-phosphomolybdate/silicone rubber - ISE, (6) *, poor selectivity	790
	(1) DP, (2) Cs tetraphenylborate-LISE, (3) pH 6—8, (4) NH_4^+, Ag^+, Hg^{2+}, (5) $> 10^{-4}M$, (6) *	791
	(1) DP, (2) $CsBPh_4$ in $PhNO_2$, (3) pH 4—12, (5) 10^{-1}—$10^{-6}M$	791a
Fe^{3+}	(1) BT(EDTA, Cu^{2+}), (2) Cu^{2+}-ISE, (5) $10^{-2} - 5 \times 10^{-5}M(0.1\ mM)$	759, 780, 791
	(1) T(EDTA), (2) Cu^{2+}-ISE, (4) acetate or hexamine buffer, pH 5.0, (5) 0.25 mM, (6) Cu-EDTA indicator	760, 780, 791
	(1) DP, (2) $Fe_nSe_{60}Ge_{28}Sb_{12}(n = 1.3$—2) -ISE, (5) 10^{-2}—$10^{-6}M$, (6) *	792, 793
	(1) DP, (2) coated-wire ISE($FeCl_4^-$. Aliquat 336 S), (4) Hg^{2+}, Fe^{2+}, Zn^{2+}, (6) *, in iron ores without separation of SiO_2	794
	(1) T(EDTA), (2) F^--ISE, (3) pH 2.2 (4) ions with log $K_{d(EDTA)}$ $> 10^{14}$, (5) 5—350 µg, (6) NaF indicator	795
	(1) AS($10^{-3}M\ F^-$), (2) F^--ISE, (3) acetate buffer, pH 4.6, (5) 10^{-1}—$10^{-4}M$	781
	(1) T(Na/NH_4pyrrolidinedithiocarbamate), (2) HgS/Hg_2Cl_2-ISE, (6) *	208
La^{3+}	(1) T(F^-), (2) F^--ISE, (6) coulometric	796
	(1) T(EDTA), (2) Cu^{2+}-ISE, (3) NH_3 buffer, pH 10, (6) Cu-EDTA indicator	760
Mn^{2+}	(1) DP, (2) $MnTe/Ag_2S$-ISE, (5) $> 10^{-6}M$, (6) *, slope ~ 25 mV/pMn	723
	(1) BT(EDTA, Cu^{2+}), (2) Cu^{2+}-ISE, (3) acetate buffer, pH 4.6	759
MnO_4^-	(1) DP, (2) ClO_4^--ISE	289
Mo	(1) AS(I^-), (2) I^--ISE, (4) Zr^{4+}, V^{4+}, Th^{4+}, Fe^{3+}, Tl^+, S^{2-}, reductants, (5) > 4 µg/l, (6) catalyt. reaction, $H_2O_2 - I^-$	797
	(1) DP, (2) $[Et_4N]_2 [MoO(SCN)_5]$-ISE, (4) Fe, V, W, Nb, Re, (5) $10^{-2} - 5 \times 10^{-8}M$, (6) *	798
Ni^{2+}	(1) AS($10^{-3}M\ CN^-$), (2) CN^--ISE, (3) pH 11.3, (5) 10^{-1}—$10^{-4}M$	781
	(1) DP, (2) NiSe (or NiTe)/Ag_2S-ISE, (6) *	723
	(1) DP, (2) bis-(O,O'-di-isobutyldithiophosphate)nickel(II)-ISE, (5) 10^{-1}—$10^{-4}M$, (6) *	799
	(1) DP, (2) various poorly selective ISE's, (6) *	800, 801

continued

Table 4.12 — cont.

Determinand	Procedure	Ref.
	(1) T(o-phen or EDTA), (2) Cu^{2+}-ISE, (3) neutral, (5) $10^{-3}M$, (6) Cu-EDTA indicator	675
Ni^{2+}	(1) BT(EDTA, Cu^{2+}), (2) Cu^{2+}-ISE, (3) acetate buffer, pH 4.75	759
	(1) T(Na/NH_4pyrrolidinedithiocarbamate), (2) HgS/Hg_2Cl_2-ISE, (6) *	208
Pd^{2+}	(1) DP, (2) quarternary ammonium ion-$[PdCl_4]^{2-}$-LISE, (6) *	802
Rb^+	(1) DP, (2) valinomycin K^+-ISE, (4) K^+, Cs^+	611
	(1) DP, (2) univalent glass-ISE, (4) univalent ions	688
ReO_4^-	(1) DP, (2) ClO_4^--ISE	290
	(1) DP, (2) Brilliant Green-ReO_4^--LISE, (3) pH 5—7.2, (5) 10^{-2}—$10^{-5}M$	802a
Sc^{3+}	(1) BT(EDTA, Cu^{2+}), (2) Cu^{2+}-ISE, (5) 0.5—5 mg	803
Sm	(1) T(EDTA), (2) Cu^{2+}-ISE, (3) acetate buffer, pH 5, (6) Cu-EDTA indicator	760
Sr^{2+}	(1) DP, (2) neutral ion-carrier-ISE, (4) Ba^{2+}, Cs^+, (6) *	804
	(1) BT(Na_2SO_4, Pb^{2+}), (2) Pb^{2+}-ISE, (3) mixed	425
Tl^+	(1) DP, (2) TlI/AgI-ISE, (6) *	806
	(1) DP, (2) O,O'-didecyldithiophosphate-LISE, (3) pH 5—12	806a
	(1) DP, (2) univalent glass-ISE	556, 805
Th^{4+}	(1) BT(EDTA, Cu^{2+}), (2) Cu^{2+}-ISE, (5) 0.5—5 mg	803
	(1) DP, (2) Th^{4+}-dinonylnaphthalenesulphonate-ISE, (6) *, poorly selective	807
	(1) DP, (2) F^--ISE, (3) NaOAc-TEA-HCl buffer, pH 5.5, (5) $> 2 \times 10^{-5}M$, (6) F^- displaced from ThF_4 measured	808
	(1) T(EDTA), (2) Cu^{2+}-ISE, (3) acetate buffer, pH 5.0, (6) Cu-EDTA indicator	760
UO_2^{2+}	(1) DP, (2) UO_2-ester of organophosphoric acid in PVC, (3) pH 3, (4) Fe^{3+}, $> 10^{-3}M$ NO_3^-, ClO_4^-, (5) $> 10^{-4}M$, (6) *	809
	(1) DP, (2) Methylene Blue-ISE, (3) excess of benzoate to complex U(VI), (6) *, response to tribenzoatodioxouranate (VI)	810
	(1) DP, (2) F^--ISE, (3) $(NH_4)_2CO_3$-TEA-HCl, pH 5.5, (5) $> 10^{-4}M$, (6) F^- displaced from UO_2F_2 measured	808
	(1) DP, (2) UO_2^{2+}-tetracyanoquinodimethane	808a
W	(1) AS(I^-), (2) I^--ISE, (4) large amts. Mo(VI), Zr^{4+}, Th^{4+}, Fe^{3+}, Tl^+, S^{2-}, reductants, (5) > 4 μg/l, (6) catalyt. reaction, H_2O_2 — I^-	797
Zn^{2+}	(1) DP, (2) $ZnSe/Ag_2S$-ISE, (5) $> 10^{-5}M$, (6) *	723
	(1) DP, (2) Brilliant Green tetracyanozincate-ISE, (3) 20 times	812

continued

Table 4.12 — cont.

Determinand	Procedure	Ref.
	SCN^-, (5) $> 10^{-5} M$, (6) *	
	(1) DP, (2) quarternary NH_4^+-salt-$[ZnCl_4]^{2-}$-LISE, (6) *	802
	(1) $T(S^{2-})$, (2) CuS/Ag_2S-ISE, (5) 0.2—10 μg	811
	(1) T(o-phen or EDTA or TEPA), (2) Cu^{2+}-ISE, (3) pH 10 (NH_4OH), (6) Cu-L indicator	675
	(1) BT(EDTA, Cu^{2+}), (2) Cu^{2+}-ISE, (3) acetate buffer, pH 4.75, (5) $10^{-2} - 5 \times 10^{-5} M$	759
	(1) T(Na/NH_4pyrrolidinedithiocarbamate), (2) HgS/Hg_2Cl_2-ISE, (6) *	208
	(1) T(tri- or tetraethylenepentamine), (2) Ag_2S-ISE, (3) borate buffer, pH 9—10, (4) no interference from Ca^{2+}, Mg^{2+}, Al^{3+}, Fe^{3+}, (5) 0.02—2 mM, (6) $AgNO_3$ indicator	813
	(1) DP, (2) Zn^{2+}-di-n-octylphenylphosphate-ISE (4) Ca^{2+}, (5) 10^{-2}—$10^{-5} M$	813a
Zr^{4+}	(1) T(EDTA), (2) Cu^{2+}-ISE, (3) acetate buffer, pH 5.0, (6) Cu-EDTA indicator	760
	(1) DP, (2) F^--ISE, (3) Na salicylate-TEA—HCl, pH 5.5, (6) F^- liberated from ZrF_4 measured	808
total oxidants in the air	(1) DP, (2) I^--ISE, (3) absorption solution, (5) $< 10^{-4} M$	814

4.20 DETERMINATION OF VARIOUS ORGANIC COMPOUNDS

Determinations not discussed in previous sections are summarized in Table 4.13.

Table 4.13

Determination of various organic materials

(Abbreviations: BT — back-titration (in brackets first the reagent, then the back-titrant), DP — direct potentiometry, T — titration (titrant in brackets), * — laboratory made electrode).

Order of entries: (1) method, (2) electrode, (3) medium, (4) interferences, (5) concentration range, (6) other remarks.

Determinand	Procedure	Ref.
acetylcholine	(1) DP, (2) acetylcholinetetra(p-chlorophenyl)borate-ISE, (4) no interference from Na^+, NH_4^+, K^+, (5) 10^{-1} to $10^{-5}M$, (6) *.	815
oxalate, benzoate, formate, acetate, propionate, m-toluate etc.	(1) DP, (2) salts of Aliquat 336 S, (6) *	218, 285
oxalate	(1) T($CaCl_2$), (2) Ca^{2+}-ISE, (3) pH 7—11	816
salicylate, biphthalate, 1-naphthalene-sulphonate	(1) DP, (2) quarternary NH_4^+-salts in PVC, (6) *	817
salicylate	(1) DP, (2) quarternary NH_4^+-salicylate, (6) *	818
rhodanese enzyme	(1) DP, (2) CN^--ISE, (6) enzymatic reaction monitored	819
dodecylsulphate anion	(1) DP, (2) $C_{12}H_{25}SO_4^- \cdot C_{16}H_{33}(CH_3)_3N^+$-ISE, (6) *	820
aromatic sulphonates	(1) DP, (2) Crystal Violet-aromatic sulphonate-ISE, (6) *	821
p-toluene sulphonate (laurylbenzene sulphonate)	(1) DP, (2) salts of sulphonate + Aliquat 336 S-ISE, (6) *, critical micelle conc. can be monitored; rather poor selectivity	822
alkylbenzene sulphonate	(1) DP, (2) sulphonate-ferroin-PVC-ISE, (6) *	823
surfactants	(1) DP, (2) 40 % w/w hexadecyltrimethylammonium dodecylsulphate(silicone rubber)-ISE, (6) *	824, 825
8-quinolinol-5-sulphonate	(1) DP, (2) 8-quinolinol-5-sulphonate, zephiramine-ISE, (6) *	826
vicinal glycols	(1) DP, (2) ClO_4^--ISE, (3) acetate buffer, pH 4 + IO_4^-, (6) residual concentration of IO_4^- monitored after oxidation of glycol	838

continued

Table 4.13 — cont.

Determinand	Procedure	Ref.
glucose	(1) DP, (2) I^--ISE, (6) enzymatic reaction, glucose $+ O_2 \rightarrow$ gluconic acid $+ H_2O_2$ followed by reaction $H_2O_2 + 2I^- + 2H^+ \rightarrow 2H_2O + I_2$	827
penicillin	(1) DP, (2) pH-penicillase enzyme ISE, (3) pH 6.4, (5) 5×10^{-2}—$10^{-4}M$, (6) *	828, 829
EDTA	(1) T (Cu^{2+}), (2) Cu^{2+}-ISE	675
epoxides	(1) BT (HBr, Ag^+), (3) acetonitrile	830
choline	(1) DP, (2) cholinetetrakis(p-chlorophenyl)borate or cholinetetrakis(p-phenoxyphenyl)borate-ISE, (6) *	831, 832
picrate	(1) DP, (2) n-tetrahexylammonium picrate-ISE, (6) *	833
maleic and phthalic acid	(1) DP, (2) Crystal Violet or tris (bathophenanthroline) Fe(II) + acid anion-ISE, (6) *	834
phthalate isomers	(1) DP, (2) NO_3^--ISE, (6) after liquid chromatographic separation	834a
trifluoroacetate	(1) DP, (2) Crystal Violet + acid anion-LISE, (4) ClO_4^-, SCN^-, I^-; no interference from bivalent ions, Cl^-, F^-, OAc^-, (5) 10^{-1}—$3 \times 10^{-5}M$, (6) *	835
formaldehyde	(1) T(Ag^+), (2) I^--ISE	836
cholesterol	(1) DP, (6) *, double enzyme procedure, automated	837
saccharine	(1) DP, (2) subst. phenanthroline + saccharine-ISE, (5) 10 μM — $0.1M$, (6) *	839

REFERENCES

1. M. S. Frant: *US Patent*, 3 431 182 (4. 2. 1966).
2. M. S. Frant and J. W. Ross, Jr.: *Science*, **154**, 1553 (1966).
3. A. M. G. Macdonald and K. Tóth: *Anal. Chim. Acta*, **41**, 99 (1968).
4. J. Veselý: *Chem. Listy*, **65**, 86 (1971).
5. J. Veselý and K. Štulík: *Anal. Chim. Acta*, **73**, 157 (1974).
6. N. V. Bausova, V. G. Bamburov, L. I. Manakova and A. P. Sivoplyas: *Zh. Analit. Khim.*, **28**, 2042 (1973).
7. A. Koller: Private communication.
8. Perkin—Elmer Corp.: *Brit. Patent*, 1 240 028 (5. 8. 1968); 1 298 719 (24. 4. 1969).
9. R. A. Durst and J. K. Taylor: *Anal. Chem.*, **39**, 1483 (1967).
10. R. A. Durst (ed.): *Ion-Selective Electrodes*, NBS No. 314, Washington (1969).
11. P. Venkateswarlu: *Clin. Chim. Acta*, **59**, 277 (1975).
11a. A. Hallsworth, J. A. Weatherell and D. Deutsch: *Anal. Chem.* **48**, 1660 (1976).

12. A. Sher, R. Solomon, K. Lee and M. W. Muller: *Phys. Rev.*, **144**, 593 (1966).
12a. O. O. Lyalin and M. S. Turaeva: *Elektrokhimiya*, **13**, 256 (1977).
13. J. Kummer and M. E. Milberg: *Chem. Eng. News*, **47**, 90 (1969).
14. R. R. Tarasyants, R. N. Potoepkina, V. P. Roze and E. A. Bondarevskaya: *Zh. Analit. Khim.*, **27**, 808 (1972).
15. R. Bock and S. Strecker: *Z. Anal. Chem.*, **235**, 322 (1968).
16. T. B. Warner: *Anal. Chem.*, **41**, 527 (1969).
17. J. W. Ross, Jr.: Chapter 2 in ref. 10.
18. I. Sekerka and J. F. Lechner: *Talanta*, **20**, 1167 (1973).
19. T. Anfält and D. Jagner: *Anal. Chim. Acta*, **47**, 483 (1969); **50**, 23 (1970).
20. J. E. Harwood: *Water Res.*, **3**, 273 (1969).
21. E. W. Baumann: *Anal. Chim. Acta*, **54**, 189 (1971).
22. J. Buffle, N. Parthasarathy and W. Haerdi: *Anal. Chim. Acta*, **68**, 253 (1973).
23. J. Koryta: *Ion-Selective Electrodes*, Cambridge Univ. Press, Cambridge (1975).
24. J. Veselý: *J. Electroanal. Chem.*, **41**, 134 (1973).
25. N. Parthasarathy, J. Buffle and D. Monnier: *Anal. Chim. Acta*, **68**, 185 (1974).
26. E. Heckel and P. F. Marsh: *Anal. Chem.*, **44**, 2347 (1972).
26a. S. Tanikava, T. Adachi, N. Shiraishi, G. Nakagawa and K. Kodama: *Bunseki Kagaku*, **25**, 646 (1976).
27. J. J. Lingane: *Anal. Chem.*, **40**, 935 (1968).
28. R. A. Robinson, W. C. Duer and R. G. Bates: *Anal. Chem.*, **43**, 1862 (1971).
29. J. Knoeck: *Anal. Chem.*, **41**, 2069 (1969).
30. B. Norén: *Acta Chem. Scand.*, **23**, 931 (1969).
31. E. W. Baumann: *Anal. Chim. Acta*, **42**, 127 (1968).
32. K. Srinivasan and G. A. Rechnitz: *Anal. Chem.*, **40**, 509 (1968).
33. J. N. Butler: Chapter 5 in ref. 10.
34. H. Pokorná: *Thesis*, Charles University, Prague 1975.
35. A. Vaillant, J. Devynck and B. Trémillon: *J. Electroanal. Chem.*, **57**, 219 (1974).
36. L. J. Warren: *Anal. Chim. Acta*, **53**, 199 (1971).
37. H. H. Broene and T. deVries: *J. Am. Chem. Soc.*, **69**, 1644 (1947).
38. P. A. Evans, G. J. Moody and J. D. R. Thomas: *Lab. Pract.*, **20**, 644 (1971).
39. G. J. Moody and J. D. R. Thomas: *Proc. Soc. Anal. Chem.*, **8**, 84 (1971).
40. M. J. D. Brand and G. A. Rechnitz: *Anal. Chem.*, **42**, 478 (1970).
41. C. R. Edmond: *Anal. Chem.*, **41**, 1327 (1969).
42. M. Frant and J. W. Ross, Jr.: *Anal. Chem.*, **40**, 1169 (1968).
43. T. B. Warner: *Water Res.*, **5**, 459 (1971).
44. M. A. Peters and D. M. Ladd: *Talanta*, **18**, 655 (1971).
45. N. Shiraishi, Y. Murata, G. Nakagawa and K. Kodama: *Bunseki Kagaku*, **23**, 176 (1974).
46. P. Grametbauer and J. Vebr: Private communication.
47. Kuo-Hsing Chang, Shih-Yian Yin, Hou-chu Wong: *Huaxue Tongbao* (*Chung. Bull.*), 36 (1975).
48. S. Tanikawa, H. Kirihara, N. Shiraishi, G. Nakagawa and K. Kodama: *Anal. Letters*, **8**, 879 (1975).
49. J. J. Lingane: *Anal. Chem.*, **39**, 881 (1967).
50. T. Anfält, D. Dyrssen and D. Jagner: *Anal. Chim. Acta*, **43**, 487 (1968).
51. T. S. Light, R. F. Mannion and K. S. Fletcher: *Talanta*, **16**, 1141 (1969).
52. T. S. Light and R. F. Mannion: *Anal. Chem.*, **41**, 107 (1969).
53. T. Eriksson and G. Johansson: *Anal. Chim. Acta*, **52**, 465 (1970).

54. G. Harzdorf: *Z. Anal. Chem.*, **245**, 67 (1969).
55. D. J. Curran and K. S. Fletcher: *Anal. Chem.*, **41**, 267 (1969).
56. G. Muto and K. Nozaki: *Bunseki Kagaku*, **18**, 247 (1969).
57. T. B. Warner and D. J. Bressan: *Anal. Chim. Acta*, **63**, 165 (1973).
58. D. Weiss: *Chem. Listy*, **63**, 1153 (1969).
59. P. J. Ke and L. W. Regier: *Anal. Chim. Acta*, **53**, 23 (1971).
60. H. Chermett, C. Martelet, D. Sandino, M. Benmalek and J. Tousset: *Anal. Chim. Acta*, **59**, 373 (1972).
61. H. H. Willard and O. B. Winter: *Ind. Eng. Chem., Anal. Ed.*, **1**, 5 (1933).
62. K. E. McLeod and H. L. Crist: *Anal. Chem.*, **45**, 1272 (1973).
63. Z. Kubec and E. Maierová: *Chem. Průmysl*, **21**, 46, 388 (1971).
64. A. Hrabéczy-Páll, K. Tóth, E. Pungor and F. Vallo: *Anal. Chim. Acta*, **77**, 278 (1975).
65. *Orion Res. Newsletter*, **III** (Nos, 3, 4) (1970).
66. L. A. Elfers and C. E. Decker: *Anal. Chem.*, **40**, 1658 (1968).
67. T. Okida, K. Kaneda, T. Yanaka and R. Sugai: *Atm. Environ.*, **8**, 927 (1974).
68. K. Svoboda and H. Ixfeld: *Staub-Reinhalt*, **31**, 1 (1971).
68a. M. Mascini: *Anal. Chim. Acta*, **85**, 287 (1976).
69. J. C. van Loon: *Anal. Letters*, **1**, 393 (1968).
70. T. A. Palmer: *Talanta*, **19**, 1141 (1972).
71. J. C. Bast: *Chem. Z.*, **96**, 108 (1972).
72. W. L. Plüger and G. H. Friedrich: *Proc. 4th Intern. Geochem. Explor. Symp., London*, p. 421 (1972).
72a. M. O. Schwartz and G. H. Friedrich: *J. Geochem. Expl.*, **2**, 103 (1973).
72b. S. Tanikawa, H. Kirihara, N. Shiraishi, S. Nakagawa and K. Kodama: *Bunseki Kagaku*, **24**, 559 (1975).
73. B. L. Ingram: *Anal. Chem.*, **42**, 1825 (1970).
74. H. Bösch and H. Weingerl: *Z. Anal. Chem.*, **262**, 104 (1972).
75. R. L. Clements, G. A. Sargeant and P. J. Webb: *Analyst*, **96**, 51 (1971).
76. R. T. Oliver and A. G. Clayton: *Anal. Chim. Acta*, **51**, 409 (1970).
77. W. H. Ficklin: *Prof. Paper US Geol. Survey*, **700C**, C 186 (1970).
78. R. C. Harris and H. H. Williams: *J. Appl. Meteorol.*, **8**, 299 (1969).
79. D. E. Collis and A. A. Diggens: *Water Treat. Exam.*, **18**, 192 (1969).
80. R. S. Nanda and K. Kapoor: *Ind. J. Med. Res.*, **60**, 949 (1972).
81. *Orion Res., Appl. Bull.* 5A.
82. N. T. Crosby: *J. Appl. Chem.*, **19**, 100 (1969).
83. Orion Research: *Anal. Meth. Guide*, 7th Ed., May (1975).
84. E. Belack: *Am. Water Works Assoc.*, **64**, 62 (1972).
85. F. Brudevold, E. Moreno and Y. Bakhos: *Arch. Oral. Biol.*, **17**, 1155 (1972).
86. F. Schöller and D. W. Kasper: *Gas(Wasser)Wärme*, **24**, 115 (1970).
87. P. G. Brewer, D. W. Spencer and F. E. Wilkins: *Deep-Sea Res.*, **17**, (1970).
88. H. L. Windom: *Limn. Oceanogr.*, **16**, 806 (1971).
89. T. B. Warner: *Geophys. Res.*, **77**, 2728 (1972).
90. P. Chambon, R. Chambon and J. Vial: *Trib. CEBEDEAU*, **27**, 432 (1974).
91. W. C. Hanson and D. J. Lloyd: *Chem. Ind. (London)*, **41**, 1971 (1972).
92. Z. Sosin: *Chem. Anal.*, **18**, 1005 (1973).
93. Z. Kubec: *Chem. Průmysl*, **21**/46, 564 (1971).
94. E. F. Croomes and R. C. McNutt: *Analyst*, **93**, 729 (1968).
95. *Orion Res. Appl. Bull.* A4.

96. D. E. Jordan: *J. Assoc. Off. Anal. Chem.*, **53**, 447 (1970).
97. *Orion Res. Appl. Bull.* A2.
98. M. S. Frant: *Plating*, **54**, 702 (1967).
99. *Orion Res. Appl. Bull.* A1.
100. T. Erickson: *Anal. Chim. Acta*, **65**, 417 (1973).
101. H. H. Moecken, H. Eschrich and G. Willerborts: *Anal. Chim. Acta*, **45**, 233 (1969).
102. Y. Veno and M. Tsuiki: *Denki Kagaku*, **38**, 278 (1970).
103. K. Oshima and N. Shibata: *Bunseki Kagaku*, **23**, 392 (1974).
104. L. G. Bruton: *Anal. Chem.*, **43**, 579 (1971).
105. P. J. Ke and L. W. Regier: *J. Fish Res.*, **28**, 1055 (1971).
106. S. Larsen and A. E. Widdowson: *J. Soil Sci.*, **22**, 210 (1971).
107. B. L. Farell: *J. Geoch. Expl. Amsterdam*, 3 (1974).
108. W. W. Wood: *J. Res. US Geol. Survey*, **1**, 237 (1973).
109. T. Ishii and M. Ihida: *Bunseki Kagaku*, **22**, 665 (1973).
110. S. H. Sherken and J. C. Williams: *Interbur. By-Lines*, **6**, 225 (1970).
111. B. A. Raby and W. E. Sunderland: *Anal. Chem.*, **39**, 1304 (1968).
112. U. Westerlund-Helmerson: *Anal. Chem.*, **43**, 1120 (1971).
113. F. Tobias: *Aluminium* (*Düsseldorf*), **48**, 431 (1972).
114. C. E. Plucinski: *US At. Energy Comm.* BNWL 601 (1968).
115. J. Thomas and H. J. Gluskoter: *Anal. Chem.*, **46**, 1321 (1974).
116. E. Hepp: *Apothekerf. Tidds.*, **80**, 327 (1972).
117. R. Pouget: *Chim. Anal.* (*Paris*), **53**, 479 (1971).
118. M. Verloo and A. Cottenie: *Med. Fac. Landb. Univ. Ghent*, **34**, 137 (1969).
119. E. J. Duff and J. L. Stuart: *Talanta*, **19**, 76 (1972).
120. E. J. Duff and J. L. Stuart: *Anal. Chim. Acta*, **52**, 155 (1970).
121. J. Tušl: *Chem. Listy*, **64**, 322 (1970).
122. L. Evans, R. D. Hoyle and J. B. Macashill: *New Zealand J. Sci.*, **13**, 143 (1970).
123. E. J. Duff and J. L. Stuart: *Anal. Chim. Acta*, **57**, 233 (1971).
124. T. Kojima, M. Ichise and Y. Seo: *Talanta*, **19**, 539 (1972).
125. R. A. Powell and M. C. Stokes: *Atm. Environ.*, **7**, 169 (1973).
126. R. S. Yungham and T. B. McMullen: *Fluoride*, **3**, 143 (1970).
127. A. Liberti and M. Mascini: *Fluoride*, **4**, 49 (1971).
128. J. Thompson, T. B. McMullen and G. B. Morgan: *J. Air Pollut. Contr. Assoc.*, **21**, 484 (1971).
129. *Orion Res. Newsletter*, II, Nos. 7, 8 (1970).
130. J. Tušl: *Anal. Chem.*, **44**, 1693 (1972).
131. B. W. Fry and D. R. Taves: *J. Lab. Clin. Med.*, **75**, 1020 (1970).
132. D. R. Taves: *Nature*, **217**, 1051 (1968).
133. L. Singer and W. D. Armstrong: *Anal. Chem.*, **40**, 613 (1968).
134. A. Zober and B. Schellmann: *Z. Klin. Chem. Klin. Biochem.*, **13**, 197 (1975).
135. R. C. Rittner and T. S. Ma: *Mikrochim. Acta*, 404 (1972).
136. C. W. Louw and J. F. Richards: *Analyst*, **97**, 334 (1972).
137. W. Schöniger: *Mikrochim. Acta*, 123 (1955); 869 (1956).
138. W. Schöniger: *Z. Anal. Chem.*, **150**, 306 (1956); **154**, 158 (1957).
139. W. J. Kirsten: *Mikrochem. J.*, **7**, 34 (1963).
140. B. Schreiber and R. W. Frei: *Mikrochim. Acta*, 219 (1975I).
141. J. N. Wilson and C. Z. Marczewski: *Anal. Chem.*, **45**, 2409 (1973).
142. P. J. Ke, L. W. Regier and H. E. Power: *Anal. Chem.*, **41**, 1081 (1969).
143. J. Pavel, R. Kuebler and H. Wagner: *Mikrochem. J.*, **15**, 192 (1970).

144. W. I. Rogers and J. A. Wilson: *Anal. Biochem.*, **32**, 31 (1969).
145. W. Selig: *Mikrochim. Acta*, 337 (1970).
146. W. Selig: *Z. Anal. Chem.*, **249**, 30 (1970).
147. M. B. Terry and F. Kasler: *Mikrochim. Acta*, 569 (1971).
148. D. A. Shearer and G. F. Morris: *Mikrochem. J.*, **15**, 199 (1970).
149. L. Helešic: *Collection Czech. Chem. Commun.*, 37, 1514 (1972).
150. H. J. Francis, J. H. Deonarine and D. D. Persing: *Mikrochem. J.*, **14**, 580 (1969).
151. M. E. Aberlin and C. A. Bunton: *J. Org. Chem.*, 35, 1825 (1970).
152. B. C. Jones, J. E. Heveran and B. Z. Senkowski: *J. Pharm. Sci.*, **60**, 1036 (1971).
153. D. R. Taves: *Talanta*, 15, 1015 (1968).
154. P. Venkateswarlu: *Anal. Chem.*, **46**, 878 (1974).
155. B. F. Erlanger and R. A. Sack: *Anal. Biochem.*, **33**, 318 (1970).
156. H. G. McCann: *Arch. Oral Biol.*, 13, 475 (1968).
157. B. Woodward, N. F. Taylor and R. V. Brunt: *Anal. Biochem.*, **36**, 303 (1970).
158. E. J. Bushee, D. K. Grisson and D. R. Smith: *J. Dent. Child.*, **38**, 279 (1971).
159. B. C. Jones, J. E. Heveran and B. Z. Senkowski: *J. Pharm. Sci.*, **58**, 607 (1969).
160. R. L. Baker: *Anal. Chem.*, **44**, 1326 (1972).
161. J. S. Jacobson and I. L. Heller: *J. Assoc. Off. Anal. Chem.*, **58**, 1129 (1975).
162. J. S. Jacobson and D. C. McCune: *J. Assoc. Off. Anal. Chem.*, **55**, 991 (1972).
163. J. R. Melton, W. L. Hoover and J. L. Ayers: *J. Assoc. Off. Anal. Chem.*, **57**, 508 (1974).
164. L. Torma and B. E. Ginther: *J. Assoc. Off. Anal. Chem.*, **51**, 1181 (1968).
165. H. E. Power and L. W. Regier: *J. Sci. Food Agric.*, **21**, 108 (1970).
166. W. P. Ferren and N. A. Shane: *J. Food Sci.*, **34**, 317 (1969).
167. G. DeBaenst, J. Mertens, P. van den Winkel and D. L. Massart: *J. Pharm. Belg.*, **28**, 188 (1973).
168. C. Martin and S. Brun: *Trav. Soc. Pharm. Montpell.*, **29**, 161 (1969).
169. J. Jongeling, T. D. Flissenbaalje and I. Gedalia: *J. Dent. Res.*, 588 (1973).
170. W. J. Simpson and J. Tuba: *J. Oral Med.*, **23**, 104 (1968).
171. C. Fuchs, D. Dorn, C. A. Fuchs, H. V. Henning, C. McIntosh, F. Scheeler and M. Stennert: *Clin. Chim. Acta*, **60**, 157 (1975).
172. J. C. Jardillier and G. Desmet: *Clin. Chim. Acta*, **47**, 357 (1973).
173. L. Singer and W. D. Armstrong: *Arch. Oral Biol.*, **14**, 1343 (1969).
174. D. Taves: *J. Dent. Res.*, **50**, 783 (1971).
175. P. Venkateswarlu, L. Singer and W. D. Armstrong: *Anal. Biochem.*, **42**, 350 (1971).
176. P. Grøn, H. G. McCann and F. Brudevold: *Arch. Oral Biol.*, **13**, 203 (1968).
177. F. Brudevold, Y. Bakhos and P. Grøn: *Arch. Oral Biol.*, **18**, 699 (1973).
178. I. L. Shannon, R. P. Suddick and E. J. Edmonds: *Car. Res.*, **7**, 1 (1973).
179. H. J. Wespi and W. Burgi: *Gynaecol.*, **168**, 443 (1969).
180. *Orion Res. Newsletter*, III (Nos. 9, 10) (1971).
181. L. Singer, W. D. Armstrong and J. J. Vogel: *J. Lab. Clin. Med.*, **74**, 354 (1969).
182. H. J. Wespi and W. Burgi: *Car. Res.*, 5, 89 (1971).
183. Y. Ericsson, I. Hellström and Y. Hofvander: *Acta Paediat. Scand.*, **61**, 459 (1972).
184. G. Bang, T. Kristoffersen and K. Meyer: *Acta Path. Microbiol. Scand.*, A **78** 49 (1970).
185. A. M. Pochomis and F. D. Griffith: *Am. Ind. Hyg. Ass. J.*, **32**, 557 (1971).
186. M. J. Larsen, M. Kold and F. R. van der Fehr: *Car. Res.*, **6**, 193 (1972).

187. H. G. McCann and P. Grøn: *Arch. Oral Biol.*, **13**, 877 (1968).
188. P. Grøn, K. Yao and M. Spinelli: *J. Dent. Res. Suppl.*, **48**, 709 (1969).
189. R. Aasenden, E. C. Moreno and F. Brudevold: *Arch. Oral. Biol.*, **17**, 355 (1972)
190. P. Hotz: *Helv. Odont. Acta*, **16**, 32 (1972).
191. C. Martin and B. Mandron: *Trav. Soc. Pharm. Montpell.*, **33**, 229 (1973).
192. N. Shane and P. Miele: *J. Pharm. Sci.*, **57**, 1260 (1968).
193. F. M. Parkins: *J. Dent. Res.*, **51**, 1346 (1972).
194. A. L. Schick: *J. Assoc. Off. Anal. Chem.*, **56**, 798 (1973).
195. R. Behrend: *Z. Physik. Chem.*, **11**, 466 (1893).
196. M. Stern, H. Schwachman, T. Licht and A. J. deBethune: *Anal. Chem.*, **30**, 1506 (1958).
197. J. Číhalík: *Potenciometrie*, Academia, Prague (1961).
198. D. J. G. Ives and G. J. Janz (eds.): *Reference Electrodes*, Academic Press, New York (1961).
199. C. Harzdorf: *Z. Anal. Chem.*, **270**, 23 (1974).
200. I. M. Kolthoff and H. L. Sanders: *J. Am. Chem. Soc.*, **59**, 416 (1937).
201. J. Veselý: *Thesis*, Charles University, Prague (1973).
201a. I. Sekerka and J. F. Lechner: *Anal. Letters.* **9**, 1099 (1976).
202. J. Veselý and J. Jindra: *Proc. IMEKO Symp. Electrochem. Sensors, Veszprém, Hungary*, p. 69 (1968); *Chem. Abstr.* **72**, 27534a (1970).
203. H. Adametzová and R. Vaďura: *J. Electroanal. Chem.*, **55**, 53 (1974).
204. G. W. S. van Osch, J. van Honwelingen and A. M. H. Weelink: *US Patent* 3 824 169 (15. 1. 1973).
204a. G. W. S. van Osch and B. Griepink: *Z. Anal. Chem.*, **273**, 271 (1975).
205. E. M. Skobets and G. A. Kleibs: *Zh. Obshch. Khim.*, **10**, 1612 (1940).
206. Orion Res.: *Brit. Patent* 1 150 698 (2. 10. 1967).
207. J. Růžička and C. G. Lamm: *Anal. Chim. Acta*, **54**, 1 (1971).
208. J. F. Lechner and I. Sekerka: *J. Electronal. Chem.*, **57**, 317 (1974).
208a. I. Sekerka, J. F. Lechner and R. Wales: *Water Res.*, **9**, 663 (1975).
209. J. Tacussel: *French Patent*, 2 203 518 (13. 10. 1972).
210. E. Pungor, J. Havas and K. Tóth: *Acta Chim. Acad. Sci. Hung.*, **41**, 239 (1964).
211. G. A. Rechnitz and M. R. Kresz: *Anal. Chem.*, **38**, 1786 (1966).
212. E. Pungor: *Anal. Chem.*, **39**, No. 13, 28A (1967).
213. M. Mascini and A. Liberti: *Anal. Chim. Acta*, **47**, 339 (1969).
214. J. C. van Loon: *Anal. Chim. Acta*, **54**, 23 (1972).
215. J. C. van Loon: *Intern. J. Environ. Anal. Chem.*, **3**, 53 (1973).
216. A. Tateda: *Mem. Fac. Sci. Kyushu Univ. Ser. C.*, **8**, 213 (1973); *Chem. Abstr.* **79**, 37964z. (1973).
217. W. M. Wise: *US Patent*, 3 801 486 (18. 5. 1972).
218. C. J. Coetzee and H. Freiser: *Anal. Chem.*, **41**, 1128 (1969).
219. J. C. Frost: *Anal. Chim. Acta*, **48**, 321 (1969).
220. J. D. Czaban and G. A. Rechnitz: *Anal. Chem.*, **45**, 471 (1973).
220a. W. M. Armstrong, W. Wojtkowski and R. W. Bixenman: *Biochim. Biophys. Acta*, **465**, 165 (1977).
220b. L. S. Kelday, D. J. F. Bowling and M. G. Penny: *J. Exp. Bot.*, **28**, 31 (1977).
221. *Handbook of Chemistry and Physics*, 51st Ed., The Chemical Rubber Co., (1970 –1971).
221a. P. K. C. Tseng and W. F. Gutknecht: *Anal. Chem.*, **48**, 1996 (1976).
222. N. A. Kazaryan, A. P. Kreshkov and T. M. Syrykh: *Zh. Fiz. Khim.*, **47**, 2590 (1973).

223. J. Nedoma: *Chem. Listy*, **65**, 71 (1971).
224. O. J. Jensen and B. Nicolaisen: Report cited in ref. 203.
225. W. Jaenicke: *Croat. Chem. Acta*, **44**, 157 (1972).
226. R. Matejec, H. D. Meissner and E. Moisar: in *Progress in Surface and Membrane Science* (J. F. Danielli, M. D. Rosenberg and D. A. Cadenhead, eds.), Vol. 6, p. 1. Academic Press, New York (1973).
227. H. Gerischer and H. Selzle: *Electrochim. Acta*, **18**, 799 (1973).
228. F. C. Brown: *Phys. Chem. Solids*, **4**, 206 (1958).
229. M. Heřman and D. Weiss: *Silikáty*, **2**, 125 (1973).
230. L. G. Sillén and A. E. Martell: *Stability Constants of Metal-Ion Complexes*, 2nd Ed., The Chemical Society, London (1964).
230a. H. A. Klasens and J. Goossen: *Anal. Chim. Acta*, **88**, 41 (1977).
231. E. J. Duff and J. L. Stuart: *Analyst*, **100**, 739 (1975).
232. D. Weiss: *Chem. Listy*, **65**, 305 (1971).
233. J. Havas, E. Papp and E. Pungor: *Acta Chim. Acad. Sci. Hung.*, **58**, 9 (1968).
234. E. J. Duff and J. L. Stuart: *Chem. Ind. (London)*, 1115 (1973).
234a. J. Koryta: *Anal. Chim. Acta*, **91**, 1 (1977).
235. I. M. Kolthoff and N. H. Furman: *Potentiometric Titration*, Wiley, New York (1949).
236. T. S. Prokopov: *Anal. Chem.*, **42**, 1096 (1970).
237. S. M. Hassan: *Z. Anal. Chem.*, **266**, 272 (1973).
238. W. Krijgsman, J. F. Mansveld and B. Griepink: *Z. Anal. Chem.*, **249**, 368 (1970).
239. K. S. Fletcher and R. F. Mannion: *Anal. Chem.*, **42**, 285 (1970).
240. E. Votoček: *Chem. Ztg.*, **42**, 257 (1918).
241. Z. Fehér, G. Nagy, K. Tóth and E. Pungor: *Analyst*, **99**, 699 (1974).
242. J. C. van Loon: *Analyst*, **93**, 788 (1968).
243. S. J. Haynes and A. H. Clark: *Econ. Geol.*, **67**, 378 (1972).
244. T. G. Lee: *Anal. Chem.*, **41**, 391 (1969).
245. D. E. Erdmann: *Environ. Sci. Technol.*, **9**, 252 (1975).
246. Orion Research; Inc.: *Instr. Man. Model* 417-*Skin Chloride Measuring System*.
246a. P. T. Bray, G. C. F. Clark, G. J. Moody and J. D. R. Thomas: *Clin. Chim. Acta*, **77**, 69 (1977).
247. E. Pungor and K. Tóth: *Analyst*, **95**, 625 (1970).
248. D. A. Katz and A. K. Mukherji: *Microchem. J.*, **13**, 604 (1968).
249. J. Papp: *Sv. Papp. Tidn.*, **75**, 677 (1972).
250. M. Marounek and V. Ulrych: *Listy Cukrov.*, **89**, 155 (1973).
251. *Orion Res. Newsletter:* II, Nos. 5, 6 (1970).
251a. K. Fukamachi and N. Ishibashi: *Bunseki Kagaku*, **26**, 69 (1977).
252. A. R. Selmer-Olsen and A. Øien: *Analyst*, **98**, 412 (1973).
253. K. Torrance: *Analyst*, **99**, 203 (1974).
254. W. J. van Oort, L. Gipon, C. Bognar, J. C. van Heyst, P. C. M. Frintrop and B. Griepink: *Z. Anal. Chem.*, **270**, 200 (1974).
255. N. Ogata: *Bunseki Kagaku*, **21**, 780 (1972).
256. A. Lerman and A. Shatkay: *Earth Planet Sci. Lett.*, **5**, 63 (1968).
257. B. W. Hipp and G. W. Langdale: *Commun. Soil Sci. Plant. Anal.*, **2**, 237 (1971).
258. I. Torok: *Agr. Talaj*, **21**, 394 (1972).
259. C. G. Lamm, E. H. Hansen and J. Růžička: *Anal. Letters*, **5**, 451 (1972).
260. R. M. Speights, J. D. Brooks and A. J. Barnard: *J. Pharm. Sci.*, **60**, 748 (1971).

261. M. S. Frant: *Plating*, **58**, 686 (1971).
262. Y. M. Dessouky, K. Tóth and E. Pungor: *Analyst*, **95**, 1027 (1970).
263. Z. Vajda and J. Kovacs: *Hung. Sci. Instrum.*, **20**, 31 (1971).
264. I. Simonyi and I. Kálmán: in *Ion-Selective Electrodes* (E. Pungor, ed.), p. 253 Akadémiai Kiadó, Budapest (1973).
265. L. Boulares-Poinsignon, J. L. Adamy and P. Federlin: *Compt. Rend.*, *C*, **268** 1894 (1969).
266. G. Lemahien, C. Lemahien-Hode and B. Resibois: *Analusis*, **1**, 110 (1972).
267. W. Potman and E. A. M. F. Dahmen: *Mikrochim. Acta*, 303 (1972).
268. J. Kálmán, K. Tóth and D. Kuttal: *Acta Pharm. Hung.*, **41**, 267 (1971).
269. B. György, L. André, E. Stehli and E. Pungor: *Anal. Chim. Acta*, **46**, 318 (1969).
270. H. Jacin: *Die Stärke*, **25**, 271 (1973).
271. L. R. LaCroix, D. R. Keeney and L. M. Walsh: *Commun. Soil Sci. Plant Anal.*, **1**, **1** (1970).
272. V. H. Holsinger, L. P. Pozati and M. J. Pallansch: *J. Dairy Sci.*, **50**, 1189 (1967).
273. A. W. Randell and P. M. Linklater: *Aust. J. Dairy Technol.*, **27**, 51 (1972).
274. M. Kapel and J. C. Fry: *Analyst*, **99**, 608 (1974).
275. L. Kopito and H. Schwachman: *Pediatrics*, **43**, 794 (1969).
276. L. Hansen, M. Buechele, J. Koroschec and W. J. Warwick: *Am. J. Clin. Path.*, **49**, 834 (1968).
276a. G. A. Cherian and G. J. Hills: *Clin. Chem.*, **17**, 652 (1971).
277. J. D. Kruse-Jarres and G. N. Noeldge: *Aerztl. Lab.*, **21**, 259 (1975).
278. W. Krijgsman, J. F. Mansveld and B. Griepink: *Clin. Chim. Acta*, **29**, 575 (1970).
279. H. J. Degenhart, G. Abein, B. Bevaart and J. Baks: *Clin. Chim. Acta*, **38**, 217 (1972).
280. E. Papp and E. Pungor: *Z. Anal. Chem.*, **246**, 26 (1969).
281. H. Dahms, R. Rock and D. Salingson: *Clin. Chem.*, **14**, 859 (1969).
282. J. Havas: *Hung. Sci. Instr.*, **27**, 47 (1974).
283. N. Ishibashi and H. Kohara: *Anal. Letters*, **4**, 785 (1971).
284. R. J. Rohm and G. G. Guilbault: *Anal. Chem.*, **46**, 590 (1974).
285. H. J. James, G. Carmack and H. Freiser: *Anal. Chem.*, **44**, 856 (1972).
286. A. G. Fogg, A. S. Pathan and D. T. Burns: *Anal. Chim. Acta*, **73**, 220 (1974).
287. M. Sharp: *Anal. Chim. Acta*, **65**, 405 (1973).
288. R. E. Reinsfelder and F. A. Schultz: *Anal. Chim. Acta*, **65**, 425 (1973).
289. R. J. Baczuk and R. J. DuBois: *Anal. Chem.*, **40**, 685 (1968).
290. R. F. Hirsch and J. D. Portock: *Anal. Letters*, **2**, 295 (1969).
291. M. Kataoka and T. Kambara: *J. Electroanal. Chem.*, **73**, 279 (1976).
292. K. Srinivasan and G. A. Rechnitz: *Anal. Chem.*, **41**, 1203 (1969).
293. J. Vošta and J. Havel: *Scripta Chem.*, **3**, 99 (1973).
294. M. P. Henderson, V. I. Miasek and T. W. Swaddle: *Can. J. Chem.*, **49**, 317 (1971).
295. G. M. Shean and K. Sollner: *Ann. N. Y. Acad. Sci.*, **137**, 759 (1966).
296. M. J. Smith and S. E. Manahan: *Anal. Chim. Acta*, **48**, 315 (1969).
297. I. Sekerka and J. F. Lechner: *J. Electroanal. Chem.*, **69**, 339 (1976).
297a. P. K. C. Tseng and W. F. Gutknecht: *Anal. Letters*, **9**, 795 (1976).
298. R. Galli and T. Mussini: *Italian Patent* 903 290 (27. 4. 1970).
299. V. M. Krasnoperov and L. N. Moskvin: *Zh. Fiz. Khim.*, **48**, 3052 (1974).
300. E. M. Skobets, L. I. Makovetskaya and Yu. P. Makovetski: *Zh. Analit. Khim.*, **29**, 845 (1974).
301. J. E. Burroughs and A. I. Attia: *Anal. Chem.*, **40**, 2052 (1968).

302. Y. Israel and L. Gonen: *Israel J. Chem.*, **9**, 15 (1971).

303. S. Poser, W. Poser and B. Müller-Oerlinghausen: *Z. Klin. Chem. Klin. Biochem.*, **12**, 350 (1974).

303a. J. Angerer: *J. Clin. Chem. Clin. Biochem.*, **15**, 201 (1977).

304. W. H. Ficklin and W. C. Gotschall: *Anal. Letters*, **6**, 617 (1973).

305. D. L. Turner: *J. Food Sci.*, **37**, 791 (1972).

306. R. Wawro and G. A. Rechnitz: *Anal. Chem.*, **46**, 806 (1974).

307. G. A. Rechnitz, M. R. Kresz and S. B. Zamochnick: *Anal. Chem.*, **38**, 973 (1966).

308. J. Janošová, J. Šenký̌r and M. Bartušek: *Chem. Listy*, **67**, 836 (1973).

309. N. Ishibashi, H. Kohara and N. Uemura: *Bunseki Kagaku*, **21**, 1072 (1972).

310. Y. Shijo: *Bull. Chem. Soc. Japan*, **48**, 1647 (1975).

311. J. Růžička and K. Rald: *Anal. Chim. Acta*, **53**, 1 (1971).

312. M. Novkirishka and R. Christova: *Anal. Chim. Acta*, **78**, 63 (1975).

313. E. Pungor and E. Hollós-Rokosinyi: *Acta Chim. Acad. Sci. Hung.*, **27**, 63 (1961).

314. H. Malissa, M. Grasserbauer, E. Pungor, K. Tóth, K. Pápay and L. Pólos: *Anal. Chim. Acta*, **80**, 223 (1975).

315. H. Malissa and G. Jellinek: *Z. Anal. Chem.*, **245**, 70 (1969).

316. W. A. Bassett and T. Takahashi: *Am. Miner.*, **50**, 1576 (1965).

317. R. Despotovič, Z. Despotovič, M. Jajetič, M. Mirnik, S. Popovič and Ž. Telišman: *Croat. Chem. Acta*, **42**, 445 (1970).

318. J. Veselý: *Collection Czech. Chem. Commun.*, **39**, 710 (1974).

319. Landolt-Börnstein: *Zahlenwerte und Funktionen*, II Band, 2. Teil, p. 234, Springer, Berlin (1960).

320. M. N. Hull and A. A. Pilla: *J. Electrochem. Soc.*, **118**, 72 (1972).

321. W. E. Morf, G. Kahr and W. Simon: *Anal. Chem.*, **46**, 1538 (1974).

322. D. J. Crombie, G. J. Moody and J. D. R. Thomas: *Anal. Chim. Acta*, **80**, 1 (1975).

322a. J. Kontoyannakos, G. J. Moody and J. D. R. Thomas: *Anal. Chim. Acta*, **85**, 47 (1976).

323. R. P. Buck and V. R. Shepard, Jr.: *Anal. Chem.*, **46**, 2097 (1974).

324. W. A. B. Donners and D. A. de Wooys: *J. Electroanal. Chem.*, **52**, 277 (1974).

325. J. H. Kennedy and E. Boodman: *J. Phys. Chem.*, **74**, 2174 (1970).

326. I. Sekerka and J. F. Lechner: *Water Res.*, 10, 479 (1976).

327. D. Weiss: *Chem. Listy*, **65**, 1091 (1971).

327a. F. G. K. Baucke: *Z. Anal. Chem.*, **282**, 105 (1976).

328. M. H. Sorrentino and G. A. Rechnitz: *Anal. Chem.*, **46**, 943 (1974).

328a. H. Malissa, M. Grasserbauer, E. Pungor, K. Tóth, M. K. Papay and L. Pólos: *Anal. Chim. Acta*, **80**, 223 (1975).

329. D. Weiss: *Chem. Listy*, **66**, 858 (1972).

330. J. Havas, M. Huber, I. Szabó and E. Pungor: *Hung. Sci. Instrum.*, **9**, 19 (1967).

331. B. Paletta: *Mikrochim. Acta*, 1210 (1969).

332. H. Arino and H. H. Kramer: *Nucl. App.*, **48**, 356 (1968).

333. B. Paletta and K. Panzenbeck: *Clin. Chim. Acta*, **26**, 11 (1969).

334. E. D. Moorhead and P. W. The: *J. Appl. Chem. Biotechnol.*, **22**, 441 (1972).

335. J. H. Woodson and H. A. Liebhafsky: *Anal. Chem.*, **41**, 1894 (1969).

336. R. A. Hasty: *Mikrochim. Acta*, 925 (1973).

337. S. Honda, K. Sudo, K. Karebi and K. Tariura: *Anal. Chim. Acta*, **77**, 274 (1975).

338. R. A. Carter: *Proc. Assoc. Clin. Biochem.*, **5**, 67 (1968).

339. T. Braun. C. Ruiz de Pardo and E. C. Salazar: *Radiochem. Radioanal. Letters*, **3**, 419 (1970).

340. W. L. Hoover, J. R. Melton and P. A. Howard: *J. Assoc. Off. Anal. Chem.*, **54** 760 (1971).

341. L. L. Gerchman and G. A. Rechnitz: *Z. Anal. Chem.*, **230**, 265 (1967).

342. K. Tóth and E. Pungor: *Anal. Chim. Acta*, **51**, 221 (1970).

342a. C. Harzdorf: *Anal. Chim. Acta*, **86**, 103 (1976).

343. B. Fleet and H. von Storp: *Anal. Letters*, **4**, 425 (1971).

344. J. Veselý, O. J. Jensen and B. Nicolaisen: *Anal. Chim. Acta*, **62**, 1 (1972).

345. M. S. Frant, J. W. Ross, Jr., and J. H. Riseman: *Anal. Chem.*, **44**, 2227 (1972).

345a. H. Clysters, F. Adams and F. Verbeek: *Anal. Chim. Acta*, **83**, 27 (1976).

346. A. M. Azzam and I. A. W. Shimi: *Z. Anorg. Chem.*, **321**, 284 (1963).

347. G. P. Bound, B. Fleet, H. v. Storp and D. H. Evans: *Anal. Chem.*, **45**, 788 (1973).

348. B. Fleet and H. von Storp: *Anal. Chem.*, **43**, 1575 (1971).

349. *Orion Res. Newsletter:* **VI**, No. 1 (1974).

350. J. H. Riseman: *Am. Lab.*, **4**, 63 (1972).

351. J. W. Ross, Jr., J. H. Riseman and J. A. Krueger: *Pure Appl. Chem.*, **36**, 473 (1973)

351a. M. S. Zetter: *BRD Patent* 2 646 314 (28. 10. 1975).

352. M. Mascini: *Anal. Chem.*, **45**, 614 (1973).

353. M. Mascini and A. Napoli: *Anal. Chem.*, **46**, 447 (1974).

354. P. D. Goulden, B. K. Afghan and P. Brooksbank: *Anal. Chem.*, **44**, 1845 (1972).

355. J. L. Penland and G. Fische: *Metallobrfl. Angew. Elektrochem.*, **10**, 391 (1972).

356. B. Fleet and A. Y. W. Ho: *Talanta*, **20**, 793 (1973).

357. F. Oehme: *Chem. Techn.*, **3**, 27 (1974).

358. *Orion Res. Newsletter*, **V**, No. 1 (1973).

359. L. N. Lapatnick: *Anal. Chim. Acta*, **72**, 430 (1974).

360. G. Hermann: *GFR Patent* 2 064 822 (1972).

361. H. F. Wasgestian: *Z. Anal. Chem.*, **246**, 237 (1969).

362. W. J. Blaedel, D. B. Easty, L. Anderson and T. R. Farrell: *Anal. Chem.*, **43**, 890 (1971).

363. J. T. Gilingham, M. M. Stirer and N. R. Page: *Agron. J.*, **61**, 717 (1969).

364. T. S. Light and J. L. Swartz: *Anal. Letters*, **1**, 825 (1968).

365. C. Collombel, J. P. Durand, J. Bureau and J. Cotte: *J. Eur. Toxicol.*, **3**, 291 (1970).

366. D. G. Vickroy and G. L. Gaunt, Jr.: *Tobacco*, **174**, 50 (1972).

367. Y. M. Dessouky and E. Pungor: *Analyst*, **96**, 442 (1971).

367a. M. E. Hofton: *Environ. Sci. Technol.* **10**, 277 (1976) .

367b. P. J. Cusbert: *Anal. Chim. Acta*, **87**, 429 (1976).

368. R. A. Llenado and G. A. Rechnitz: *Anal. Chem.*, **43**, 1457 (1971).

368a. M. Mascini and A. Liberti: *IUPAC Intern. Symp. on Ion-Selective Electrodes*, Paper No. 3,Cardiff 1973.

369. Orion Res.: *Brit. Patent* 1 222 476 (6. 9. 1968).

370. M. Mascini: *Anal. Chim. Acta*, **62**, 29 (1972).

371. N. Ishibashi and K. Kina: *Bull. Chem. Soc. Japan*, **46**, 2454 (1973).

372. H. Komiya: *Bunseki Kagaku*, **21**, 911 (1972).

373. G. Nota: *Anal. Chem.*, **47**, 763 (1975).

374. P. Szarvas, I. Korondan and M. Szabó: *Magy. Kém. Foly.*, **80**, 207 (1974).

374a. M. Bartušek and J. Šenkýř: *Scr. Fac. Sci. Nat. Univ. Purkynianae Brun.*, **5**, 61 (1975); *Chem. Abstr.*, **85**, 84716 (1976).

375. M. S. Frant and J. W. Ross, Jr.: *BRD Patent* 1 598 453 (6. 12. 1965).

376. E. Pungor, E. Schmidt and K. Tóth: *Proc. IMEKO Symp. on Electrochem. Sensors,* Veszprém, Hungary, p. 121 (1968).
377. M. Mascini and A. Liberti: *Anal. Chim. Acta,* **51,** 231 (1970).
378. T. M. Hseu and G. A. Rechnitz: *Anal. Chem.,* **40,** 1054 (1968).
379. V. A. Kremer and M. A. Zarechenskii: *Kontr. Tekhnol. Protsessov Obagashch. Polez.* (Skop. No. 3, 34 (1971); *Chem. Abstr.* **78,** 78923h. (1973).
380. D. Weiss: *Chem. Listy,* **68,** 528 (1971).
381. E. W. Baumann: *Anal. Chem.,* **46,** 1345 (1974).
382. R. Bock and H. J. Puff: *Z. Anal. Chem.* **240,** 381 (1968).
383. M. Koebel: *Anal. Chem.* **46,** 1559 (1974).
384. C. H. Lin and S. Shen: *Anal. Chem.,* **36,** 1652 (1964).
385. R. Naumann and C. Weber: *Z. Anal. Chem.,* **253,** 111 (1971).
386. J. Slanina, E. Buysman, J. Agterdenbos and B. Griepink: *Mikrochim. Acta,* 657 (1971).
387. J. Papp and J. Havas: *Hung. Sci. Instrum.,* **17,** 17 (1970).
388. J. Papp: *Cellul. Chem. Technol.,* **5,** 147 (1971).
389. S. Ikeda, H. Satake, T. Hisano and T. Terazawa: *Talanta,* **19,** 1650 (1972).
390. P. Hollis: *Soil. Sci.,* **114,** 456 (1972).
391. E. J. Green and D. Schnither: *Marine Chem.,* **2,** 111 (1974).
392. D. Jagner and K. Åren: *Anal. Chim. Acta,* **52,** 491 (1970).
393. B. Fleet and A. Y. W. Ho: *Anal. Chem.,* **46,** 9 (1974).
394. J. M. Murphy and G. A. Sergeant: *Analyst,* **99,** 515 (1974).
395. B. Poljáková and A. Boček: *Hutn. Listy,* **24,** 360 (1969).
396. C. P. Morie: *Tobacco,* **173,** 34 (1971).
397. J. Slanina, P. Vermeer, J. Agterdenbos and B. Griepink: *Mikrochim. Acta,* 607 (1973).
398. M. A. Volodina and G. A. Martinova: *Zh. Analit. Khim.,* **26,** 1002 (1971).
399. W. C. Harris, E. P. Crowell and D. H. McMahon: *Tappi,* **57,** 82 (1974).
400. Orion Research: *Appl. Bull.* No. 15 (1970).
401. F. Peter and R. Rosset: *Anal. Chim. Acta,* **64,** 397 (1973).
402. L. C. Gruen and B. S. Harrap: *Anal. Biochem.,* **42,** 377 (1971).
403. W. Selig: *Mikrochim. Acta,* 453 (1973).
404. M. K. Pápay, K. Tóth and E. Pungor: *Anal. Chim. Acta,* **56,** 291 (1971).
405. M. T. Neshkova, V. P. Izvekov, M. K. Pápay, K. Tóth and E. Pungor: *Anal. Chim. Acta,* **75,** 439 (1975).
406. F. Oehme: *Chem.-Tech.,* **4,** 183 (1975).
407. T. B. Warner: *Mar. Techn. Soc. J.,* **6,** 24 (1972).
408. A. E. Mor, V. Scotta, G. Marcenaro and G. Alabeso: *Anal. Chim. Acta,* **75,** 159 (1975).
409. T. S. Light: *Ind. Water Eng.,* **6,** 33 (1969).
410. G. Spath, H. J. Niefind and M. Martina: *Monatschr. Breuer.,* **4,** 91 (1972).
411. E. F. Mellon and H. A. Gruber: *J. Am. Leath. Chem. Ass.,* **65,** 154 (1970)
412. J. L. Swartz and T. S. Light: *Tappi,* **53,** 90 (1970).
413. M. K. Pápay, V. P. Izvekov, K. Tóth and E. Pungor: *Anal. Chim. Acta,* **69,** 173 (1974).
414. D. W. Gruenwedel and R. K. Patnaik: *Agr. Food Chem.,* **19,** 775 (1971).
415. E. Rolia and J. C. Ingles: *Can. Mining J.,* **92,** 94, 97, 102 (1971).
416. B. S. Harrab and L. C. Gruen: *Anal. Biochem.,* **42,** 398 (1971).
417. A. S. Buchanan and E. Heymann: *J. Colloid. Sci.,* **4,** 151 (1949).

418. E. Pungor, J. Havas and K. Tóth: *Z. Chem.*, **5**, 9 (1965).

419. R. B. Fischer and B. F. Babcock: *Anal. Chem.*, **30**, 1732 (1958).

419a. A. M. Saunders: *US Patent* 3 709 811 (21. 9. 1970).

420. D. J. G. Ives and F. R. Smith: p. 393 in ref. 198.

421. M. S. Mohan and G. A. Rechnitz: *Anal. Chem.*, **45**, 1323 (1973).

422. N. Akimoto and K. Hozumi: *Anal. Chem.*, **46**, 766 (1974).

423. R. J. Levins: *Anal. Chem.*, **43**, 1045 (1971); **44**, 1544 (1972).

423a. M. J. Y. Jaber, G. J. Moody and J. D. R. Thomas: *Analyst*, **101**, 179 1976).

424. R. Jasinski and I. Trachtenberg: *Anal. Chem.*, **44**, 2373 (1972).

425. C. Harzdorf: *Z. Anal. Chem.*, **262**, 167 (1972).

426. J. W. Ross, Jr. and M. S. Frant: *Anal. Chem.*, **41**, 967 (1969).

427. J. O. Goertzen and J. D. Oster: *Soil Sci. Soc. Am. Proc.*, **36**, 691 (1972).

428. W. Selig and A. Salomon: *Mikrochim. Acta*, 663 (1974).

429. M. Mascini: *Analyst*, **98**, 325 (1973).

430. W. Selig: *Mikrochim. Acta*, 168 (1970).

431. R. N. Heistand and C. T. Blake: *Mikrochim. Acta*, 212 (1972).

432. J. E. Hicks, J. E. Eleenor and H. R. Smith: *Anal. Chim. Acta*, **68**, 480 (1974).

432a. H. Clysters and F. Adams: *Anal. Chim. Acta*, **92**, 251 (1977).

433. L. B. Bjoerkevoll, S. Reite and B. Salvesen: *Meddr. Norsk. Farm. Selsk.*, **37**, 72 (1975).

434. Electronic Instruments: *Leaflet* No. 33-0009-001, (December 1974).

435. J. A. Krueger: *Anal. Chem.*, **46**, 1339 (1974).

436. L. Reiszner and P. West: *Environ. Sci. Technol.*, **7**, 526 (1973).

436a. M. Mascini and T. Muratori: *Ann. Chim. (Rome)*, **65**, 287 (1975).

437. M. Young, J. N. Driscoll and K. Mahoney: *Anal. Chem.*, **45**, 2283 (1973).

438. Orion Research: *Instr. Manual SO$_2$-Electrode Model 95-64*.

439. E. H. Hansen, H. B. Filbo and J. Růžička: *Anal. Chim. Acta*, **71**, 225 (1974).

440. G. Eisenman (ed.): *Glass Electrodes for Hydrogen and Other Cations*, Dekker, New York (1967).

441. G. I. Goodfellow and H. M. Webber: *Analyst*, **97**, 95 (1972).

442. R. P. Scholer and W. Simon: *Chimia*, **24**, 372 (1970).

443. Z. Štefanac and W. Simon: *Chimia*, **20**, 436 (1966).

444. J. Mertens, P. van den Winkel and D. L. Massart: *Anal. Letters* **6**, 81 (1973).

445. E. F. Vansant, H. Deelstra and R. Dewolfs: *Talanta*, **21**, 608 (1972).

446. G. A. Rechnitz and G. Kugler: *Z. Anal. Chem.*, **210**, 174 (1965).

447. J. Růžička, E. H. Hansen, P. Bisgaard and E. Reymann: *Anal. Chim. Acta*, **72**, 215 (1974).

448. E. H. Hansen and N. R. Larsen: *Anal. Chim. Acta*, **78**, 459 (1975).

449. J. Růžička and E. H. Hansen, *Anal. Chim. Acta*, **69**, 129 (1974).

450. T. Anfält, A. Granelli and D. Jagner: *Anal. Chim. Acta*, **76**, 253 (1975).

451. R. F. Thomas and R. L. Booth: *Environ. Sci. Technol.*, **7**, 523 (1973).

452. T. R. Gilbert and A. M. Clay: *Anal. Chem.*, **45**, 1757 (1973).

453. D. Midgley and K. Torrance: *Analyst*, **97**, 626 (1972).

454. Orion Research: *Instr. Manual, Ammonia Electrode Model 95-10* (1974).

455. P. J. Leblanc and J. F. Slivinski: *Am. Lab.*, **5**, 51 (1973).

456. W. H. Evans and D. F. Partridge: *Analyst*, **99**, 367 (1974).

457. L. Vandevenne and J. Ondewater: *Trib. CEBEDEAU*, **26**, 127 (1973).

458. J. Barica and J. Fisheris: *Res. Bd. Canada*, **30**, 1389 (1973).

459. J. M. Bremner and M. A. Tabatabai: *Comm. Soil Sci. Plant Anal.*, **3**, 159 (1972).

460. W. L. Banwart, M. A. Tabatabai and J. M. Bremner: *Comm. Soil Sci. Plant Anal.*, **3**, 470 (1972).

460a. A. R. Mazoyer: *Ann. Agron.*, **23**, 673 (1972).

461. M. L. Eagan and L. Dubois: *Anal. Chim. Acta*, **70**, 157 (1974).

462. D. Rogers and K. H. Pool: *Anal. Letters*, **6**, 801 (1973).

463. J. H. C. Hoge, H. J. A. Hazenberg and C. H. Gips: *Clin. Chim. Acta*, **55**, 273 (1974).

464. F. Enning, H. J. Proelss and B. W. Wright: *Clin. Chem.*, **19**, 1162 (1973).

465. N. J. Park and J. C. B. Fenton: *J. Clin. Pathol.*, **26**, 802 (1973).

466. E. Byrne and T. Power: *Comm. Soil Sci. Plant Anal.*, **5**, 51 (1974).

467. D. J. McWilliam and C. S. Ough: *Am. J. Enol. Viticult.*, **25**, 67 (1974).

468. T. J. Wisk and K. J. Siebert: *Proc. Am. Soc. Brew. Chem.*, 26 (1973).

469. A. R. Deschreider and R. Meaux: *Analusis*, **2**, 442 (1973).

470. C. M. Dozinel: *Schweiz. Landwirtsch. Forsch.*, **12**, 301 (1973).

471. S. van Wasselaer and F. Crucke: *Ann. Pharm. Fr.*, **31**, 769 (1973).

472. S. A. Katz and G. A. Rechnitz: *Z. Anal. Chem.*, **196**, 248 (1963).

473. R. A. Llenado and G. A. Rechnitz: *Anal. Chem.*, **46**, 1109 (1974).

474. G. G. Guilbault and J. G. Montalvo, Jr.: *J. Am. Chem. Soc.*, **92**, 2533 (1970).

475. G. G. Guilbault and E. Hrabánková: *Anal. Chim. Acta*, **52**, 287 (1970).

476. G. G. Guilbault and M. Tarp: *Anal. Chim. Acta*, **73**, 355 (1974).

476a. G. Johansson and L. Ögren: *Anal. Chim. Acta*, **84**, 23 (1976).

477. G. G. Guilbault and G. Nagy: *Anal. Chem.*, **45**, 417 (1973).

478. T. Anfält, A. Granelli and D. Jagner: *Anal. Letters*, **6**, 691 (1973).

479. G. G. Guilbault and W. Stokhro: *Anal. Chim. Acta*, **76**, 237 (1975).

480. T. A. Neubecker and G. A. Rechnitz: *Anal. Letters*, **5**, 653 (1972).

481. H. Thompson and G. A. Rechnitz: *Anal. Chem.*, **46**, 246 (1974).

482. G. G. Guilbault and E. Hrabánková: *Anal. Letters*, **3**, 53 (1970); *Anal. Chem.*, **42**, 1779 (1970).

483. G. G. Guilbault and G. Nagy: *Anal. Letters*, **6**, 301 (1973).

484. A. M. Berjonneau, D. Thomas and G. Broun: *Pathol. Biol.*, **22**, 497 (1974).

485. C. Calvot, A. M. Berjonneau, G. Gelf and D. Thomas: *FEBS Letters*, **59**, 258 (1975).

485a. G. Johansson, K. Edstrom and L. Ögren: *Anal. Chim. Acta*, **85**, 55 (1976).

485b. C. P. Hsiung, S. S. Kuan and G. G. Guilbault: *Anal. Chim. Acta*, **90**, 45 (1977).

485c. J. M. Tokarsky and S. B. Weiner: *Bioelectrochem. Bioenerg.*, **3**, 106 (1976).

485d. S. L. Tong and G. A. Rechnitz: *Anal. Letters*, **9**, 1 (1976).

486. M. Matsui and H. Freiser: *Anal. Letters*, **3**, 161 (1970).

487. V. P. Izvekov, M. Kucsera-Pápay, K. Tóth and E. Pungor: *Analyst*, **97**, 634 (1972).

488. R. Jasinski, G. Barna and J. Trachtenberg: *J. Electrochem. Soc.*, **121**, 1575 (1974).

489. Orion Research: *Instr. Manual, Nitrogen Oxide Electrode Model 95-46* (1974).

490. B. M. Kneebone and H. Freiser: *Anal. Chem.*, **45**, 449 (1973).

491. G. P. Morie, C. S. Ledford and C. A. Glover: *Anal. Chim. Acta*, **60**, 397 (1972).

492. D. Kuroda: *Bunseki Kagaku*, **22**, 1191 (1973).

493. R. DiMartini: *Anal. Chem.*, **42**, 1102 (1970).

494. L. A. Dee, H. H. Martens, C. I. Merrill, J. T. Nakamura and F. C. Jaye: *Anal. Chem.*, **45**, 1477 (1973).

495. M. A. Tabatabai: *Comm. Soil Sci. Plant Anal.*, **5**, 569 (1974).

496. J. W. Ross, Jr.: *Can. Patent* 816 843 (23. 2. 1968).
497. W. M. Wise: *US Patent* 3 671 413 (18. 1. 1971).
498. J. E. W. Davies, G. J. Moody and J. D. R. Thomas: *Analyst*, **97**, 87 (1972).
499. E. A. Materova, A. L. Grekovich and N. V. Garbuzova: *Zh. Analit. Khim.*, **29**, 1900 (1974).
500. T. Nomura and G. Nakagawa: *Anal. Letters*, **8**, 873 (1975).
500a. E. M. Skobets, L. I. Makovetskaya and Yu. P. Makovetskii: *Zh. Analit. Khim.*, **29**, 2354 (1974).
500b. J. Šenkýř and J. Petr: Private communication.
500c. E. Hopîrtean, E. Stefaniga and C. Liteanu: *Chem. Anal. (Warsaw)* **21**, 867 (1976).
501. A. Hulanicki, R. Lewandowski and M. Maj: *Anal. Chim. Acta*, **69**, 409 (1974).
502. D. Weiss: *Chem. Listy*, **69**, 202 (1975).
503. S. S. Potterton and W. D. Shults: *Anal. Letters*, **1** (2), 11 (1967).
504. *Orion Res. Newsletter*, II, Nos. 9, 10 (1970).
505. Orion Research: *Instr. Manual, Nitrate Ion Electrode Model 92-07* (1973).
506. P. J. Milham, A. S. Awad, R. E. Paull and J. H. Bull: *Analyst*, **95**, 751 (1970).
507. J. S. DiGregorio and M. D. Morris: *Anal. Chem.*, **42**, 99 (1970); *Anal. Letters*, **1**, 811 (1968).
507a. Orion Res.: *Instr. Manual NO_3^--ISE, Model 93-07* (1976).
508. S. J. Bourget: *Can. Soil Sci.*, **48**, 369 (1968).
509. M. K. Mahendrappa: *Soil Sci.*, **108**, 132 (1969).
510. A. Øien and A. R. Selmer-Olsen: *Analyst*, **94**, 888 (1969).
511. A. R. Mack and R. B. Sanderson: *Can. J. Soil Sci.*, **51**, 95 (1971).
512. J. Mertens, P. van den Winkel and D. L. Massart: *Anal. Chem.*, **47**, 522 (1975).
513. D. G. Gehring, W. A. Dippel and R. S. Boucher: *Anal. Chem.*, **42**, 1686 (1970).
514. L. J. Forney and J. F. McCoy: *Analyst*, **100**, 157 (1975).
515. S. E. Manahan: *Appl. Microbiol.*, **18**, 479 (1969).
516. J. L. Paul and R. M. Carlson: *J. Agr. Food Chem.*, **16**, 766 (1968).
517. D. R. Keeney, B. H. Byrnes and J. J. Genson: *Analyst*, **95**, 383 (1970).
518. D. Langmuir and R. L. Jacobson: *Environ. Sci. Technol.*, **4**, 834 (1970).
519. R. Mazder and I. Agius: *Ann. Agron.*, **23**, 673 (1972).
520. J. Mertens and D. L. Massart: *Bull. Soc. Chim. Belg.*, **82**, 431 (1973).
521. A. V. Baker, N. H. Peck, and G. E. MacDonald: *Agron. J.*, **63**, 126 (1971).
522. K. P. Louwrier and J. R. deRijk: *Z. Anal. Chem.*, **270**, 203 (1974).
523. J. M. C. Ridden, R. R. Barefoot and J. G. Roy: *Anal. Chem.*, **43**, 1109 (1971).
524. J. O. Burman and G. Johanson: *Anal. Chim. Acta*, **80**, 215 (1975).
525. P. Chalk and D. Keeney: *Nature*, **229**, 42 (1971).
526. Y. Yasumori and T. Tatsuta: *Bunseki Kagaku*, **22**, 1069 (1973).
527. P. J. Milham: *Analyst*, **95**, 758 (1970).
528. P. Voogt: *Deutsch. Lebensm. Rund.*, **65**, 196 (1969).
529. B. D. McCaslin, W. T. Franklin and M. A. Dillon: *J. Am. Soc. Sugar Beet Technol.*, **16**, 64 (1970).
530. M. Brand and J. Dubernard: *Coton Fibr. Trop.*, **27**, 411 (1972).
531. A. W. M. Sweetsur and A. G. Wilson: *Analyst*, **100**, 485 (1975).
532. H. Jacin: *Tobacco Sci.*, **14**, 28 (1970).
533. C. C. Westcott: *Food Technol.*, **25**, 709 (1971).
534. M. Nanjo, T. J. Rohm and G. G. Guilbault: *Anal. Chim. Acta*, **77**, 19 (1975).
535. I. Nagelberg, L. I. Braddock and G. J. Barbero: *Science*, **166**, 1403 (1969).
536. F. R. Shu and G. G. Guilbault: *Anal. Lett.*, **5**, 559 (1972).

536a. O. G. Takaishvili, E. P. Motsonelidze, Yu. M. Karachentseva and P. I. Lavitaya: Elektrokhimiya, **12**, 291 (1976).

536b. I. Novozamsky and W. H. van Riemsdijk: Anal. Chim. Acta, **85**, 41 (1976).

536c. R. E. van de Leest, M. N. Bfekmans and L. Heijne: BRD Patent 2 600 846 (24. 1. 1975).

536d. J. Tacussel and J. J. Fombon: Private communication.

537. Orion Res. Newsletter, III, Nos. 1, 2 (1971).

538. T. Tanaka, K. Hiiro and K. Akimori: Z. Anal. Chem., **272**, 44 (1974).

539. W. Selig: Mikrochim. Acta, 564 (1970).

540. R. W. Stow, R. F. Baer and B. F. Randall: Arch. Phys. Med. Rehabil., **38**, 646 (1957).

541. J. W. Severinghaus and A. F. Bradley: J. Appl. Physiol., **13**, 515 (1958).

542. P. Sekelj and R. B. Goldbloom: in ref. 440.

543. O. Siggaard-Andersen: The Acid-Base Status of Blood, Munksgaard, Copenhagen (1972).

544. P. Astrup, K. Jorgensen, O. Siggaard-Andersen and K. Engel: Lancet, **1**, 1035 (1960).

545. U. Fiedler, E. H. Hansen and J. Růžička: Anal. Chim. Acta, **74**, 423 (1975).

546. W. M. Wise: US Patent 3 723 281 (31. 1. 1972).

547. H. B. Herman and G. A. Rechnitz: Science, **184**, 1074 (1974); Anal. Letters, **8**, 147 (1975).

548. A. L. Grekovich, E. A. Materova and N. V. Garbuzova: Zh. Analit. Khim., **28**, 1206 (1973).

548a. G. A. Rechnitz, G. J. Nogle, M. R. Bellinger and H. Lees: Clin. Chim. Acta, **76**, 295 (1977).

549. C. R. Caflisch and N. W. Carter: Anal. Biochem., **60**, 252 (1974).

550. R. M. Carlson and J. L. Paul: Anal. Chem., **40**, 1292 (1968).

551. A. G. Fogg, A. S. Pathan and D. T. Burns: Anal. Letters, **7**, 545 (1974).

551a. C. Liteanu, E. Hopîrtean and E. Stefaniga: Rev. Roum. Chim. **22**, 653 (1977).

552. R. L. Kochen: Anal. Chim. Acta, **71**, 451 (1974).

553. R. M. Carlson and J. L. Paul: Soil Sci., **108**, 266 (1969).

554. H. Wilde: Anal. Chem., **45**, 1526 (1973).

555. P. Lanza and P. L. Buldini: Anal. Chim. Acta, **75**, 149 (1975).

556. G. Eisenman: in Advances in Analytical Chemistry and Instrumentation (C. N. Reilley, ed.), Vol. 4, Wiley, New York (1965).

557. G. Mattock and R. Uncles: Analyst, **87**, 977 (1962).

558. E. W. Baumann: Anal. Chem., **40**, 1731 (1968).

559. A. G. Koksharov, I. V. Koksharova, V. L. Volkov and A. A. Fotiev: USSR Patent 468 893 (12. 10. 1972).

560. M. Güggi, U. Fiedler, E. Pretsch and W. Simon: Anal. Letters, **8**, 857 (1975).

560a. R. C. Thomas, W. Simon and M. Oehme: Nature, **258**, 754 (1975).

561. T. Urban and I. Steiner: J. Phys. Chem., **35**, 3058 (1931).

562. B. v. Lengyel and E. Blum: Trans. Faraday Soc., **30**, 461 (1934).

563. B. P. Nikolskii and T. A. Tolmacheva: Zh. Fiz. Khim., **10**, 504 (1937).

564. M. M. Shults and T. M. Ovchinnikova: Vestn. Leningr. Univ., No. 2, 129 (1954).

565. G. Eisenman, D. O. Rudin and J. U. Casby: Science, **126**, 831 (1957).

566. G. Mattock: Chimia, **21**, 209 (1967).

567. M. F. Wilson, E. Haikala and P. Kivalo: Anal. Chim. Acta, **74**, 395 (1975); **74**, 411 (1975).

568. E. L. Eckfeldt and W. E. Proctor: *Anal. Chem.*, **43**, 332 (1971).
569. D. Hawthorn and N. J. Ray: *Analyst*, **93**, 158 (1968).
570. H. M. Webber and A. L. Wilson: *Analyst*, **94**, 209 (1969).
571. R. N. Khuri: in ref. 10.
572. D. Ammann, E. Pretsch and W. Simon: *Anal. Letters*, **7**, 23, (1974).
572a. W. Simon, E. Pretsch, D. Ammann, W. E. Morf, M. Güggi, R. Bissig and M. Kessler: *Pure Appl. Chem.*, **44**, 613 (1975).
572b. M. Güggi, M. Oehme, E. Pretsch and W. Simon: *Helv. Chim. Acta*, **59**, 2417 (1976).
572c. U. Fiedler: *Anal. Chim. Acta*, **89**, 101 (1977).
573. P. van den Winkel, J. Mertens, G. de Beaust and D. Massart: *Anal. Letters*, **5**, 567 (1972).
574. A. A. Diggens, K. Parker and H. Webber: *Analyst*, **97**, 198 (1972).
575. I. Sekerka and J. F. Lechner: *Anal. Letters*, **7**, 463 (1974).
576. L. J. Denisova: *Teploenergetika*, **11**, 37 (1973).
577. A. H. Truesdell and B. F. Jones: *Chem. Geol.*, **4**, 51 (1969).
578. F. V. Ranzen and Z. Ya. Solovieva: *Radiokhimiya*, **15**, 101 (1973).
579. C. A. Bower: *Soil Sci. Am. Proc.*, **23**, 29 (1959).
580. E. Chamberland, E. B. Doiron and J. Can: *Plant Sci.*, **53**, 233 (1973).
581. O. M. Kotenko: *Farm. Zh. (Kiev)*, **27**. 49 (1972).
582. F. V. McNerney: *J. Assoc. Off. Anal. Chem.*, **57**, 1159 (1974).
583. H. J. Marsoner and K. Hanoucourt: *Aerztl. Lab.*, **18**, 397 (1972).
584. H. D. Portnoy, E. S. Gurjian and B. Henry: *Am. J. Clin. Pathol.*, **45**, 384 (1966).
585. E. Moore and D. W. Wilson: *J. Clin. Invest.*, **42**, 293 (1963).
586. A. Pelleg and B. G. Levy: *Clin. Chem.*, **21**, 1572 (1975).
587. F. Gotoh, Y. Tazaki, H. Hamaguchi and J. S. Meyer: *J. Neurochem.*, **9**, 81 (1962).
587a. P. T. Bray *et al.*: *A Perspective of Sodium and Chloride ISE Sweat Tests for Screening in Cystic Fibrosis*, Chem. Dept. Univ. of Wales, Cardiff 1975.
588. N. Akimoto and K. Hozumi: *Bunseki Kagaku*, **21**, 1490 (1972).
589. B. L. Lenz and J. R. Mold: *Tappi*, **54**, 2051 (1971).
590. G. A. Rechnitz: *Chem. Eng. News*, **45**, No. 25, 146 (1967).
591. G. Baum and W. M. Wise: *US Patent* 3 598 713 (3. 6. 1969).
592. W. M. Wise, M. J. Kurey and G. Baum: *Clin. Chem.*, **16**, 103 (1970).
593. G. Baum and M. Lynn: *Anal. Chim. Acta*, **65**, 393 (1973).
594. J. E. W. Davies, G. J. Moody, W. M. Price and J. D. R. Thomas: *Lab. Pract.*, **22**, 20, (1973).
595. A. G. Fogg, A. S. Pathan and D. T. Burns: *Anal. Letters*, **7**, 539 (1974).
595a. P. A. Rock, T. L. Eyrich and S. Styer: *J. Electrochem. Soc.*, **124**, 530 (1977).
596. G. Eisenman, G. Szabo, S. Ciani, S. McLaughlin and S. Krasne: in ref. 226.
597. C. J. Pedersen: *J. Am. Chem. Soc.*, **89**, 7017 (1967).
598. O. Ryba and J. Petránek: *J. Electroanal. Chem.*, **44**, 425 (1973); **67**, 321 (1976).
599. O. Ryba, E. Knižáková and J. Petránek: *Collection Czech. Chem. Commun.* **38**, 497 (1973).
599a. G. A. Rechnitz and E. Eyal: *Anal. Chem.*, **44**, 370 (1972).
600. I. Štěpánová and R. Vaďura: *Chem. Listy*, **68**, 853 (1974).
601. M. Semler and H. Adametzová: *J. Electroanal. Chem.*, **56**, 155 (1974).
601a. O. Ryba and J. Petránek: *Talanta*, **23**, 158 (1976).
602. W. L. Duax, H. Hauptman, C. M. Weeks and D. A. Norton: *Science*, **176**, 911 (1972).

603. L. A. R. Pioda, V. Stankova and W. Simon: *Anal. Letters*, **2**, 665 (1969).

604. M. S. Frant and J. W. Ross, Jr.: *Science*, **167**, 987 (1970).

605. S. Lal and G. D. Christian: *Anal. Letters*, **3**, 11 (1970).

606. J. H. Boles and R. P. Buck: *Anal. Chem.*, **45**, 2057 (1973).

607. E. Eyal and G. A. Rechnitz: *Anal. Chem.*, **43**, 1090 (1971).

608. U. Fiedler and J. Růžička: *Anal. Chim. Acta*, **67**, 179 (1973).

609. Duke University: *Brit. Patent* 1 250 635 (10. 12. 1967).

610. I. H. Krull, C. A. Mask and R. E. Cosgrove: *Anal. Letters*, **3**, 43 (1970).

611. M. Mascini and F. Pallozzi: *Anal. Chim. Acta*, **73**, 375 (1974).

612. B. P. Nikolskii, E. A. Materova, A. L. Grekovich and V. E. Yurinskaya: *Zh. Analit. Khim.*, **29**, 205 (1974).

613. J. Pick, K. Tóth, E. Pungor, M. Vašák and W. Simon: *Anal. Chim. Acta*, **64**, 477 (1973).

614. M. Shporer, O. Kedem, M. Stock and R. Bloch: *Israel Patent* 39 995 (26. 7. 1972).

615. R. W. Cattrall, S. Tribuzio and H. Freiser: *Anal. Chem.*, **46**, 2223 (1974).

616. S. D. Moss, J. Janata and G. C. Johnson: *Anal. Chem.*, **47**, 2238 (1975).

616a. M. Oehme and W. Simon: *Anal. Chim. Acta*, **86**, 21 (1976).

617. W. E. Morf, G. Kahr and W. Simon: *Anal. Letters*, **7**, 9 (1974).

618. S. M. Hammond and P. A. Lambert: *J. Electroanal. Chem.*, **53**, 155 (1974).

619. S. P. Datta and J. H. Ottaway: *Biochemistry*, 2nd Ed., p. 282. Boulliere, Tindall and Cassell, London (1969).

620. M. D. Smith, M. A. Genshaw and J. Greyson: *Anal. Chem.*, **45**, 1782 (1973).

621. A. K. Covington and R. A. Robinson: *Anal. Chim. Acta*, **78**, 219 (1975).

622. R. Geyer and I. Preuss: *Z. Chem.*, **14**, 29 (1974).

623. T. Anfält and D. Jagner: *Anal. Chim. Acta*, **66**, 152 (1973).

624. H. R. Williams and G. F. Haydon: *Queensl. J. Agr. Anim. Sci.*, **30**, 113 (1973).

625. Orion Research, 93 *Series Method Manual*.

626. L. A. R. Pioda, W. Simon, H. R. Bosshard and H. C. Curtius: *Clin. Chim. Acta*, **29**, 289 (1970).

627. M. Semler and H. Adametzová: *Chem. Průmysl*, **25**/50, 377 (1975).

628. A. Hulanicki and Z. Augustowska: *Anal. Chim. Acta*, **78**, 261 (1975).

629. G. J. Moody and J. D. R. Thomas: *Selective Ion-Sensitive Electrodes*, Merrow, Watford (1971).

630. A. Hulanicki and M. Trojanowicz: *Chem. Anal. (Warsaw)*, **18**, 235 (1973).

631. K. L. Cheng and K. Cheng: *Mikrochim. Acta*, 385 (1974).

632. M. Whitfield and J. V. Leyndekkers: *Anal. Chim. Acta*, **45**, 383 (1969).

633. M. Whitfield, J. V. Leyndekkers and J. D. Kerr: *Anal. Chim. Acta*, **45**, 399 (1969).

634. M. E. Thompson: *Science*, **153**, 866 (1966).

635. D. R. Kester and R. M. Pytkowicz: *Limn. Oceanogr.*, **13**, 670 (1968).

635a. P. Kent, S. O. Bunce, R. A. Bailley and D. A. Aikens: *Anal. Biochem.*, **62**, 75 (1974).

636. R. Přibil and V. Veselý: *Talanta*, **13**, 233 (1966).

637. T. P. Hadjiioannou and D. S. Papastathopoulos: *Talanta*, **17**, 399 (1970).

638. T. Henscheid, K. Schoenrock and P. Berger: *J. Am. Soc. Sugar Beet Technol.*, **16**, 482 (1971).

639. E. A. Moya and K. L. Cheng: *Anal. Chem.*, **42**, 1669 (1970).

640. K. L. Cheng, J. C. Hung and D. H. Prager: *Microchem. J.*, **18**, 256 (1973).

641. Orion Research: *Instr. Manual, Divalent Ion-Electrode Model 92-32*.

642. I. Sekerka and J. F. Lechner: *Anal. Letters,* **7,** 399 (1974); *Talanta,* **22,** 459 (1974).
643. M. Mascini: *Anal. Chim. Acta,* **56,** 316 (1971).
644. H. J. C. Tendeloo: *J. Biol. Chem.,* **113,** 333 (1936).
645. A. Shatkay: *Anal. Chem.,* **39,** 1056 (1967); **40,** 456 (1968).
645a. J. Veselý and J. Jindra: *Czech. Patent* 163 358 (22. 7. 1968); *Chem. Abstr.,* **86,** 65086 (1977).
646. R. M. Garrels, M. Sato, M. E. Thompson and A. H. Truesdell: *Science,* **135,** 1045 (1962).
647. J. W. Ross, Jr.: *Science,* **156,** 1378 (1967); *US Patent* 3 445 365 (10. 8. 1965).
648. D. Jagner and J. P. Østergaard-Jensen: *Anal. Chim. Acta,* **80,** 9 (1975).
649. G. H. Griffiths, G. J. Moody and J. D. R. Thomas: *Analyst,* **97,** 420 (1972).
650. G. J. Moody, R. B. Oke and J. D. R. Thomas: *Analyst,* **95,** 910 (1970).
651. R. Bloch, A. Shatkay and H. A. Saroff: *Biophys. J.,* **7,** 865 (1967).
652. E. A. Materova, A. J. Grekovich and S. E. Didina: *Elektrokhimiya,* **8,** 829 (1972).
653. R. Bloch, O. Kedem, M. Shporer and E. Lobel: *Israel Patent* 39 996 (26. 7. 1972).
654. J. Růžička, E. H. Hansen and J. Chr. Tjell: *Anal. Chim. Acta,* **67,** 155 (1973).
655. J. Růžička and J. Chr. Tjell: *Brit. Patent* 2 349 299 (2. 10. 1972).
656. D. Ammann, M. Güggi, E. Pretsch and W. Simon: *Anal. Letters,* **8,** 709 (1975).
656a. M. Oehme, M. Kessler and W. Simon: *Chimia,* **30,** 204 (1976).
657. D. Ammann, E. Pretsch and W. Simon: *Anal. Letters,* **5,** 843 (1972).
658. R. W. Cattrall, D. M. Drew and I. C. Hamilton: *Anal. Chim. Acta,* **76,** 269 (1975).
659. R. W. Cattrall and D. M. Drew: *Anal. Chim. Acta,* **77,** 9 (1975).
660. F. A. Schultz, A. J. Petersen, C. A. Mask and R. P. Buck: *Science,* **162,** 267 (1968).
661. A. Ansaldi and S. I. Epstein: *Anal. Chem.,* **45,** 595 (1973).
661a. A. Hulanicki and M. Trojanowicz: in *Ion-Selective Electrodes,* (E. Pungor, ed.), Akadémiai Kiadó, Budapest (1977.)
662. P. F. Baker: *Proc. Biophys. Mol. Biol.,* **24,** 177 (1972).
663. G. R. J. Christoffersen and E. S. Johansen: *Anal. Chim. Acta,* **81,** 191 (1976).
664. J. Bagg and R. Vinen: *Anal. Chem.,* **44,** 1773 (1972).
664a. G. J. Moody and J. D. R. Thomas: in *Ion-Selective Electrodes,* (E. Pungor, ed.), Akadémiai Kiadó, Budapest (1977).
665. R. P. Buck and J. R. Sandifer: *J. Phys. Chem.,* **77,** 2122 (1973).
666. G. A. Rechnitz and Z. F. Lin: *Anal. Chem.,* **40,** 696 (1968).
667. G. K. Pagenkopf and K. A. Buell: *Water Res.,* **8,** 375 (1974).
668. R. G. Burr: *Clin. Chim. Acta,* **43,** 311 (1973).
669. R. A. Llenado: *Anal. Chem.,* **47,** 2243 (1975).
670. H. B. Collier: *Anal. Chem.,* **42,** 1443 (1970).
671. A. Hulanicki and M. Trojanowicz: *Anal. Chim. Acta,* **68,** 155 (1974).
672. A. K. Mukherji: *Anal. Chim. Acta,* **40,** 354 (1968).
673. S. L. Tackett: *Anal. Chem.,* **41,** 1703 (1969).
674. T. Anfält and D. Jagner: *Anal. Chem.,* **45,** 2412 (1973).
675. J. W. Ross, Jr. and M. S. Frant: *Anal. Chem.,* **41,** 1900 (1969).
676. M. Mascini and A. Liberti: *Anal. Chim. Acta,* **53,** 202 (1971).
677. *Orion Research Appl. Bull.* A13 (1970).
677a. J. M. van der Meer, G. den Boef and W. E. van der Linden: *Anal. Chim. Acta,* **85,** 309 (1976).

678. E. W. Moore: in ref. 10.
679. J. H. Ladenson and G. N. Bowers, Jr.: *Clin. Chem.*, **19**, 565 (1973).
680. C. Sachs: *Presse Medicale*, **78**, 1547 (1970).
681. C. Fuchs, K. Paschen, P. G. Spieckermann and G. von Westberg: *Klin. Wochschr.*, **50**, 1 (1972).
682. V. L. Subryan, M. M. Popovtzer, S. D. Parks and E. B. Reeve: *Clin. Chem.*, **18**, 1459 (1972).
683. J. Růžička and J. Chr. Tjell: *Anal. Chim. Acta*, **47**, 475 (1969).
684. *Orion Res. Appl. Bull.* A8 (1970).
685. E. A. Woolson, J. H. Axley and P. C. Kearney: *Soil Sci.*, **109**, 279 (1970).
686. D. R. Kester and R. M. Pytkowicz: *Limn. Oceanogr.*, **14**, 686 (1969).
687. D. Jagner and K. Årén: *Anal. Chim. Acta*, **57**, 185 (1971).
688. E. O. McLean, G. H. Snyder and R. E. Franklin: *Soil. Sci. Soc. Am. Proc.*, **33**, 388 (1969).
689. R. C. Knupp: *Ceram. Bull.*, **49**, 773 (1970).
690. P. J. Muldoon and B. J. Liska: *J. Dairy Sci.*, **52**, 460 (1969).
691. R. D. Allen, J. Hobbey and R. Carriere: *J. Assoc. Off. Anal. Chem.*, **51**, 1177 (1968).
692. R. W. Cummins: *Deterg. Age*, **3**, 22 (1968).
693. J. A. Blay and J. H. Ryland: *Anal. Letters*, **5**, 653 (1971).
694. H. J. Dulce and M. Hardel: *Naturwiss.*, **55**, 137 (1968).
695. W. R. Robertson: *Clin. Chim. Acta*, **24**, 149 (1969).
696. A. E. W. Mowe and G. M. Makblouf: *Gastroenterology*, **55**, 465 (1968).
697. A. M. Bourdeau, C. Sachs, V. Presle and M. Dromini: *Le Pharm. Biol.*, **67**, 527 (1970).
698. C. Sachs and A. Bourdeau: *J. Phys. (Paris)*, **62**, 21 (1970).
699. E. W. Moore, *J. Clin. Invest.*, **49**, 318 (1970).
700. A. Perkins, M. Snyder, C. Thacher and M. R. Roles: *Transfusion*, **11**, 204 (1971).
701. E. W. Moore and A. L. Blum: *J. Clin. Invest.*, **47**, 70a (1968).
702. I. Oreskes, C. Hirsch, K. S. Douglas and S. Kupfer: *Clin. Chim. Acta*, **21**, 303 (1968).
703. S. O. Hansen and L. Theodorsen: *Clin. Chim. Acta*, **31**, 119 (1971).
704. I. C. Radde, B. Höffken, D. K. Parkinson, J. Sheepers and A. Luckham: *Clin. Chem.*, **17**, 1002 (1971).
705. G. D. Barry: *Am. J. Physiol.*, **220**, 874 (1971).
706. W. G. Robertson and M. Peacock: *Clin. Chim. Acta*, **20**, 315 (1968).
707. C. Sachs, A. M. Bourdeau, S. Balsan and V. Presle: *Ann. Biol. Clin.*, **27**, 487 (1969).
708. R. S. Hattner, J. W. Johnson, D. S. Bernstein, A. Wachman and J. Brackman: *Clin. Chim. Acta*, **28**, 67 (1970).
709. H. D. Schwartz, B. C. McConville and E. F. Christopherson: *Clin. Chim. Acta*, **31**, 97 (1971).
710. C. E. Sachs and A. M. Bourdeau: *Israel J. Med. Sci.*, **7**, 717 (1971).
711. J. E. Hinkle and L. H. Cooperman: *Brit. J. Anaesth.*, **343**, 1108 (1971).
712. B. Seamonds, J. Towfighi and D. A. Arvan: *Clin. Chem.*, **18**, 155 (1972).
713. C. E. Sachs and A. M. Bourdeau: *J. Physiol. (Paris) Suppl.* **62**, 313 (1970).
714. A. D. Perris, J. F. Whitfield and P. K. Tolg: *Nature*, **219**, 527 (1968).
715. J. E. Garvin: *J. Cell. Physiol.*, **72**, 197 (1969).

716. C. Woodward and E. A. Davidson: *Proc. Natl. Acad. Sci. USA,* **60,** 201 (1968).
717. C. E. Sachs and A. M. Bourdeau: *Rev. Eur. Etud. Clin. Biol.* **17,** 218 (1972).
718. M. S. Mohan and G. A. Rechnitz: *J. Am. Chem. Soc.,* **94,** 1714 (1972).
719. R. S. Mokady: *Israel J. Chem.,* **6,** 411 (1968).
720. K. Hozumi and N. Akimoto: *Anal. Chem.,* **42,** 1312 (1970).
721. A. L. Budd: *J. Electroanal. Chem.,* **5,** 35 (1963).
722. E. A. Materova, V. V. Moiseev and S. P. Smith-Fogelevich: *Zh. Fiz. Khim.,* **33,** 893 (1959).
723. H. Hirata and K. Higashiyama: *Talanta,* **19,** 391 (1972).
724. C. Bernard, J. P. Malugani and G. Robert: *Compt. Rend. C.,* **279,** 985 (1974).
725. M. Sharp and G. Johansson: *Anal. Chim. Acta,* **54,** 13 (1971).
726. S. Lal: *Z. Anal. Chem.,* **255,** 209 (1971).
727. M. A. Wechter and H. R. Shanks: *US Patent* 3 825 482 (16. 2. 1972).
728. D. C. Müller, P. W. West and R. H. Müller: *Anal. Chem.,* **41,** 2038 (1969).
729. R. A. Durst and B. T. Duhart: *Anal. Chem.,* **42,** 1002 (1970).
730. R. A. Durst: in ref. 10.
731. C. Liteanu, I. C. Popescu and V. Ciovîrnachȝ: *Talanta,* **19,** 985 (1972).
732. F. Tobias: *Galvanotechnik,* **63,** 644 (1972).
733. J. Vrbský and J. Fogl: *Chem. Průmysl* **22**/47, 241 (1972).
734. A. Y. Iizuka, T. Yasaki, K. Ohashi, Y. Yokota, A. Matsuzawa and M. Kato: *J. Dent. Health,* **23,** 182 (1973).
735. M. Yamazato, S. Fukuda, M. Kato and T. Ishimori: *Denki Kagaku,* **41,** 789 (1973).
736. A. V. Gordievskii, A. F. Zhukov, V. S. Shterman, N. I. Savvin and Yu. I. Urusov: *Zh. Analit. Khim.,* **29,** 1414 (1974).
737. R. W. Cattrall and Chin-Poh Pui: *Anal. Chem.,* **48,** 552 (1976).
737a. G. E. Baiulescu and N. Ciocan: *Talanta,* **24,** 37 (1977).
738. R. F. Overman: *Anal. Chem.,* **43,** 616 (1971).
738a. G. E. Baiulescu and V. V. Coşofreţ: *Talanta,* **23,** 667 (1976).
739. M. S. Frant and J. W. Ross, Jr.: *BRD Patent* 1 942 379 (1968).
740. J. Veselý, J. Grégr and J. Jindra: *Czech. Patent* 143 144 (20. 6. 1969).
741. E. H. Hansen, C. G. Lamm and J. Růžička: *Anal. Chim. Acta,* **59,** 403 (1972).
742. J. Veselý: *Collection Czech. Chem. Commun.,* **36,** 3364 (1971).
743. A. F. Zhukov, A. V. Vishnyakov, Yu. I. Urusov and A. V. Gordievskii: *Zh. Analit. Khim.,* **30,** 1614 (1975).
744. H. Hirata and K. Date: *Talanta,* **17,** 883 (1970).
745. J. Pick, K. Tóth and E. Pungor: *Anal. Chim. Acta,* **61,** 169 (1972); **65,** 240 (1973).
745a. A. Hulanicki, M. Trojanowicz and M. Cichy: *Talanta,* **23,** 47 (1976).
746. R. Jasinski, I. Trachtenberg and G. Rice: *J. Electrochem. Soc.,* **121,** 363 (1974).
746a. R. W. Cattrall and Chin-Poh Pui: *Anal. Chim. Acta,* **83,** 355 (1976).
747. J. Růžička and J. Chr. Tjell: *Anal. Chim. Acta,* **49,** 346 (1970).
748. A. Burdin, J. Mesplede and M. Porthault: *Compt. Rend. C.* **276,** 65 (1973).
749. G. A. Rechnitz and Z. F. Lin: *Anal. Letters,* **1,** 23 (1967).
750. W. J. Blaedel and D. E. Dinwiddie: *Anal. Chem.,* **46,** 873 (1974).
751. R. Jasinski, I. Trachtenberg and D. Andrychuk: *Anal. Chem.,* **46,** 364 (1974).
752. W. J. Blaedel and D. E. Dinwiddie: *Anal. Chem.* **47,** 1070 (1975).
753. G. Johansson and K. Edström: *Talanta,* **19,** 1623 (1972).
754. H. Adametzová and J. Grégr: *Chem. Průmysl,* **21**/46, 506 (1971).
754a. G. B. Oglesby, W. C. Duer and F. J. Millero: *Anal. Chem.,* **49,** 877 (1977).

754b. D. Midgley: *Anal. Chim. Acta*, **87**, 19 (1976).

755. D. J. Crombie, G. J. Moody and J. D. R. Thomas: *Talanta*, **21**, 1094 (1974).

756. Radiometer a/s: Manufacturer's literature.

757. M. J. Smith and S. E. Manahan: *Anal. Chem.*, **45**, 836 (1973).

757a. M. T. Neshkova: Private communication.

758. R. Blum and H. M. Fog: *J. Electroanal. Chem.*, **34**, 485 (1972).

759. J. M. van der Meer, G. den Boef and W. E. van der Linden: *Anal. Chim. Acta*, **76**, 261 (1975).

760. E. W. Baumann and R. M. Wallace: *Anal. Chem.*, **41**, 2072 (1969).

761. G. A. Rechnitz and N. P. Kenny: *Anal. Letters*, **2**, 395 (1969).

762. M. Mascini and A. Liberti: *Anal. Chim. Acta*, **60**, 405 (1972).

763. A. V. Gordievskii, V. S. Shterman, A. Ya. Syrchenkov, N. I. Savvin, A. F. Zhukov and Yu. I. Urusov: *Zh. Analit. Khim.*, **27**, 2170 (1972).

764. E. H. Hansen and J. Růžička: *Anal. Chim. Acta*, **72**, 365 (1974).

765. H. Hirata and K. Higashiyama: *Anal. Chim. Acta*, **54**, 415 (1971); *Bull. Chem. Soc. Japan*, **44**, 2420 (1971).

766. H. Hirata and K. Higashiyama: *Anal. Chim. Acta*, **57**, 476 (1971).

767. V. Majer, J. Veselý and K. Štulík: *Anal. Letters*, **6**, 577 (1973).

768. H. Hirata and K. Date: *Anal. Chem.*, **43**, 279 (1971).

769. M. Semler and H. Pokorná: Private communication.

770. O. Tomíček and F. Poupě: *Collection Czech. Chem. Commun.*, **6**, 408 (1934).

771. E. A. Materova, V. V. Mukhovikov and M. G. Grigoryeva: *Anal. Letters*, **8**, 167 (1975).

772. *Orion Res. Newsletter*, **V**, No. 2 (1973).

773. G. A. Rechnitz and N. C. Kenny: *Anal. Letters*, **3**, 259 (1970).

774. D. C. Cormos, I. Haiduc and P. Stetin: *Rev. Chim. (Bucharest)*, **20**, 259 (1975).

775. J. Gardiner: *Water Res.*, **8**, 23 (1974).

776. H. Hirata and K. Higashiyama: *Z. Anal. Chem.*, **257**, 104 (1971).

777. M. Mascini and A. Liberti: *Anal. Chim. Acta*, **64**, 63 (1973).

777a. J. P. Deloume et al.: *French Patent* 2 268 264 (14. 4. 1974).

778. J. Růžička and E. H. Hansen: *Anal. Chim. Acta*, **63**, 115 (1973).

778a. A. W. Cattrall and Chin-Poh Pui: *Anal. Chim. Acta*, **88**, 185 (1977).

779. M. J. D. Brand, J. J. Militell and G. A. Rechnitz: *Anal. Letters*, **2**, 523 (1969).

780. L. Šůcha and M. Suchánek: *Anal. Letters*, **3**, 613 (1970).

781. F. Oehme and L. Doležalová: *Z. Anal. Chem.*, **251**, 1 (1970).

782. E. W. Baumann: *Anal. Chem.*, **42**, 110 (1970).

783. B. Jaselskis and M. K. Bandemer: *Anal. Chem.*, **41**, 855 (1969).

784. W. Selig: *Mikrochim. Acta*, 349 (1973).

785. D. Weiss: *Z. Anal. Chem.*, **262**, 28 (1972).

785a. A. G. Fogg and A. A. Al-Sibaai: *Anal. Letters*, **9**, 33 (1976).

786. B. Fleet and G. A. Rechnitz: *Anal. Chem.*, **42**, 690 (1970).

787. F. Buhl and W. Goryl: *Pr. Nauk Uniw. Slask. Katow.*, **27**, 25 (1972).

788. T. Tanaka, K. Hiiro and A. Kawahara: *Japan Patent* 75 92794 (14. 12. 1973); *Chem. Abstr.* **84**, 79350e (1976).

789. O. S. Andreeva and N. I. Danilkin: *Elektrokhimiya*, **8**, 56 (1972).

790. C. J. Coetzee and A. J. Basson: *Anal. Chim. Acta*, **56**, 321 (1971).

791. E. W. Baumann: *Anal. Chem.*, **48**, 548 (1976).

791a. C. J. Coetzee and A. J. Basson: *Anal. Chim. Acta*, **83**, 361 (1976).

792. R. Jasinski and I. Trachtenberg: *J. Electrochem. Soc.*, **120**, 1169 (1973).

226　　　　　　　　　　　　　**Applications**　　　　　　　　　　　[Ch. 4

793. C. T. Baker and I. Trachtenberg: *J. Electrochem. Soc.*, **118**, 571 (1971).
794. R. W. Cattrall and Chin-Poh Pui: *Anal. Chim. Acta*, **78**, 463 (1975); *Anal. Chem.*, **47**, 93 (1975).
795. H. Schäfer: *Z. Anal. Chem.*, **268**, 349 (1974).
796. Y. K. Lee, K. J. Whang, K. Nozaki and G. Muto: *Bunseki Kagaku*, **20**, 1441 (1971).
797. A. Altinata and B. Pekin: *Anal. Letters*, **6**, 667 (1973).
798. A. G. Fogg, J. L. Kumar and D. T. Burns: *Anal. Letters*, **7**, 629 (1974).
799. E. A. Materova, V. V. Mukhovikov and M. G. Grigorieva: *Anal. Letters*, **8**, 167 (1975).
800. E. Pungor and E. Papp: *Acta Chim. Acad. Sci. Hung.*, **66**, 19 (1970).
801. E. Buchanan, Jr. and J. L. Seago: *Anal. Chim. Acta*, **40**, 517 (1968).
802. G. Scibona, L. Mantella and P. R. Danesi: *Anal. Chem.* **42**, 844 (1970).
802a. A. G. Fogg and A. A. Al-Sibaai: *Anal. Letters*, **9**, 39 (1976).
803. L. Šůcha, M. Valentová, M. Suchánek and Z. Urner: *Scientific Papers Inst. Chem. Technol., Prague*, H9, 99 (1973).
804. E. W. Baumann: *Anal. Chem.*, **47**, 959 (1975).
805. G. Mattock: *Analyst*, **87**, 930 (1962).
806. A. V. Gordievskii, V. S. Shterman, Yu. I. Urusov, N. I. Savvin, A. Ya. Syrchenkov and A. F. Zhukova: *USSR Patent* 397 832 (21. 2. 1972).
806a. W. Szczepaniak and K. Ren: *Anal. Chim. Acta*, **82**, 37 (1976).
807. J. B. Harell, A. D. Jones and G. R. Choppin: *Anal. Chim. Acta*, **41**, 1459 (1969).
808. Fu Chung Chang, Hui-Tuh Tsai, Shaw-Chii Wu: *Anal. Chim. Acta*, **71**, 477 (1974).
808a. L. Alcacer, M. R. Barbosa, R. A. Almeida and M. F. Marzagao: *Rev. Port. Quim.*, **15**, 192 (1973).
809. D. L. Manning, J. R. Stokely and D. W. Magouyrk: *Anal. Chem.*, **46**, 1116 (1974).
810. J. R. Entwistle and T. J. Hayes: *Proc. IUPAC Symp. on Ion-Selective Electrodes*, Paper No. 17, Cardiff (1973).
811. W. J. van Oort, V. W. J. van den Bergen and B. Griepink: *Z. Anal. Chem.*, **269**, 184 (1974).
812. A. G. Fogg, M. Duzinkewycz and A. S. Pathan: *Anal. Letters*, **6**, 1101 (1973).
813. A. Hulanicki, M. Trojanowicz and S. Domanska: *Talanta*, **20**, 1117 (1973).
813a. L. Gorton and U. Fiedler: *Anal. Chim. Acta*, **90**, 233 (1977).
814. A. F. Trachtenberg and I. N. Suffet: *J. Air Pollut. Control Assoc.*, **24**, 836 (1974).
815. G. Baum: *Anal. Letters*, **3**, 105 (1970).
816. A. K. Mukherji: *Anal. Chim. Acta*, **40**, 354 (1968).
817. N. Ishibashi, A. Jyo and K. Matsumoto: *Chem. Letters*, **12**, 1297 (1973).
818. W. M. Hayes and J. H. Wagenknecht: *Anal. Letters*, **4**, 491 (1971).
819. R. A. Llenado and G. A. Rechnitz: *Anal. Chem.*, **44**, 1366 (1972).
820. B. J. Birch and D. E. Clarke: *Anal. Chim. Acta*, **61**, 159 (1972); **67**, 387 (1973).
821. N. Ishibashi, H. Kohara and K. Horinouchi: *Talanta*, **20**, 867 (1973).
822. T. Fujinaga, S. Okazaki and H. Freiser: *Anal. Chem.*, **46**, 1842 (1974).
823. T. Tanaka, K. Hiiro, A. Kawahara: *Anal. Letters*, **7**, 173 (1974).
824. A. S. Pathan: *Proc. Soc. Anal. Chem.*, **11**, 143 (1974).
825. A. G. Fogg, A. S. Pathan and D. T. Burns: *Anal. Chim. Acta*, **69**, 238 (1974).
826. M. Sugawara, T. Nakajima and T. Kambara: *J. Electroanal. Chem.*, **67**, 315 (1976).
827. G. Nagy, L. H. von Storp and G. G. Guilbault: *Anal. Chim. Acta*, **66**, 443 (1973).
828. L. F. Cullen, J. F. Rusling, A. Schleifer and G. J. Papariello: *Anal. Chem.*, **46**, 1955 (1974).

829. H. Nilsson, A. C. Akerlund and K. Mosbach: *Biochim. Biophys. Acta,* **320,** 529 (1973).

830. W. Selig and G. L. Grossman: *Z. Anal. Chem.,* **253,** 279 (1971).

831. G. Baum, M. Lynn and F. B. Ward: *Anal. Chim. Acta,* **65,** 385 (1973).

832. Corning Glass Works: *Brit. Patent* 1 304 302 (15. 6. 1970).

833. S. G. Back: *Anal. Letters,* **4,** 793 (1971).

834. A. Jyo, M. Yonemitsu and N. Ishibashi: *Bull. Chem. Soc. Japan,* **46,** 3734 (1973).

834a. F. A. Schulz and D. E. Mathis: *Anal. Chem.,* **46,** 2253 (1974).

835. N. Ishibashi and A. Jyo: *Microchem. J.,* **18,** 220 (1973).

836. S. Ikeda: *Anal. Letters,* **7,** 343 (1974).

837. D. S. Papastathopoulos and G. A. Rechnitz: *Anal. Chem.,* **47,** 1792 (1975).

838. C. H. Efstathiou and T. P. Hadjiioannou: *Anal. Chem.,* **47,** 864 (1975).

839. N. Hazemoto, N. Kamo and Y. Kobatake: *J. Assoc. Off. Anal. Chem.,* **57,** 1205 (1974).

APPENDIX

Table A

The values of the factor $(RT \ln 10)/F$ for univalent and bivalent ions
at various temperatures

°C	$i^{\pm}$ $(RT \ln 10)/F$ (V)	$i^{2\pm}$ $(RT \ln 10)/2F$ (V)	°C	$i^{\pm}$ $(RT \ln 10)/F$ (V)	$i^{2\pm}$ $(RT \ln 10)/2F$ (V)
0	0.05420	0.02710	35	0.06114	0.03057
5	0.05519	0.02760	37	0.06154	0.03077
10	0.05618	0.02809	38	0.06174	0.03087
15	0.05717	0.02859	40	0.06213	0.03107
18	0.05777	0.02889	45	0.06313	0.03157
19	0.05797	0.02899	50	0.06412	0.03206
20	0.05816	0.02908	55	0.06511	0.03256
21	0.05836	0.02918	60	0.06610	0.03305
22	0.05856	0.02928	65	0.06709	0.03355
23	0.05876	0.02938	70	0.06809	0.03405
24	0.05896	0.02948	75	0.06908	0.03454
25	0.05916	0.02958	80	0.07007	0.03504
26	0.05936	0.02968	85	0.07106	0.03553
27	0.05955	0.02978	90	0.07205	0.03603
28	0.05975	0.02988	95	0.07305	0.03653
29	0.05995	0.02998	100	0.07404	0.03702
30	0.06015	0.03008	150	0.08396	0.04694
			200	0.09388	0.04694

$R = 8.31433$ J . mole^{-1} deg^{-1}; $F = 96487$ C/eq.; $T = t°$ C $+ 273.150$; IUPAC, *Pure Appl. Chem.*, **9**, 453 (1964).

Table B

Individual activity coefficients of ions in water [taken from: J. Kielland, *J. Am. Chem. Soc.*, **59**, 1675 (1937) by permission of the American Chemical Society]

Ion	Activity coefficient					
	$I = 0.0005$	0.001	0.005	0.01	0.05	0.1
H^+	0.975	0.967	0.933	0.914	0.86	0.83
Li^+	0.975	0.965	0.929	0.907	0.835	0.80
$Rb^+, Cs^+, NH_4^+, Tl^+, Ag^+$	0.975	0.964	0.924	0.898	0.80	0.75
$K^+, Cl^-, Br^-, I^-, CN^-,$ NO_2^-, NO_3^-	0.975	0.964	0.925	0.899	0.805	0.755
$OH^-, F^-, SCN^-,$ $HS^-, ClO_3^-, ClO_4^-,$ BrO_3^-, IO_4^-, MnO_4^-	0.975	0.964	0.926	0.900	0.81	0.76
$Na^+, IO_3^-, HCO_3^-, H_2PO_4^-,$ HSO_3^-	0.975	0.964	0.928	0.902	0.82	0.775
$Hg_2^{2+}, SO_4^{2-}, S_2O_3^{2-},$ CrO_4^{2-}, HPO_4^{2-}	0.903	0.867	0.740	0.660	0.445	0.355
$Pb^{2+}, CO_3^{2-}, SO_3^{2-},$ MoO_4^{2-}	0.903	0.868	0.742	0.665	0.455	0.37
$Sr^{2+}, Ba^{2+}, Ra^{2+}, Cd^{2+},$ $Hg^{2+}, S^{2-}, WO_4^{2-}$	0.903	0.868	0.744	0.67	0.465	0.38
$Ca^{2+}, Cu^{2+}, Zn^{2+}, Sn^{2+},$ $Mn^{2+}, Fe^{2+}, Ni^{2+}, Co^{2+}$	0.905	0.870	0.749	0.675	0.485	0.405
Mg^{2+}, Be^{2+}	0.906	0.872	0.755	0.69	0.52	0.45
$PO_4^{3-}, [Fe(CN)_6]^{3-}$	0.796	0.725	0.505	0.395	0.16	0.095
$Al^{3+}, Fe^{3+}, Cr^{3+}, Sc^{3+},$ $Y^{3+}, La^{3+}, In^{3+}, Ce^{3+},$ $Pr^{3+}, Nd^{3+}, Sm^{3+}$	0.802	0.738	0.54	0.445	0.245	0.18
$[Fe(CN)_6]^{4-}$	0.668	0.57	0.31	0.20	0.048	0.021
$Th^{4+}, Zr^{4+}, Ce^{4+}, Sn^{4+}$	0.678	0.588	0.35	0.255	0.10	0.065

Table C

Standard reference values of ionic activity in solutions of sodium chloride, potassium chloride, potassium fluoride and calcium chloride at 25 °C [(R. G. Bates and R. A. Robinson: *Pure Appl. Chem.* **37**, 575 (1974) by permission of IUPAC]

molality of salt mole/kg	$-\log a_+$	$-\log a_-$	molality of salt mole/kg	$-\log a_+$	$-\log a_-$
		NaCl			KF
0.01	2.044	2.045	0.01	2.044	2.044
0.1	1.106	1.112	0.1	1.111	1.111
0.5	0.455	0.481	0.5	0.475	0.475
1.0	0.157	0.208	1.0	0.190	0.190
2.0	—0.180	—0.072	2.0	—0.119	—0.119
		KCl			CaCl$_2$
0.01	2.045	2.045	0.01	2.273	1.768
0.1	1.112	1.115	0.1	1.570	0.842
0.5	0.482	0.496	0.5	0.991	0.177
1.0	0.206	0.232	1.0	0.580	—0.140
2.0	—0.086	—0.032			

Table D

Conversion factors between ppm and mole/l.

$10^{-3}M$ equals, *ppm*		1 *ppm* equals, $10^{-5}M$	$10^{-3}M$ equals, *ppm*		1 *ppm* equals, $10^{-5}M$
26.98	aluminium	3.7	207.19	lead	0.48
17.03	ammonia	5.9	24.30	magnesium	4.1
18.04	ammonium	5.6	200.6	mercury	0.50
10.81	boron	9.3	58.71	nickel	1.7
79.90	bromide	1.3	62.0	nitrate	1.6
112.4	cadmium	0.89	99.45	perchlorate	1.0
40.08	calcium	2.5	39.1	potassium	2.6
100.09	calcium as CaCO$_3$	1.0	95.0	phosphate	1.1
			31.0	phosphorus as P	3.2
35.45	chloride	2.8			
52.0	chromium	1.9	71.0	phosphorus as (P$_2$O$_5$)/2	1.4
58.93	cobalt	1.7			
63.54	copper	1.6	107.9	silver	0.93
26.01	cyanide	3.8	23.0	sodium	4.4
19.0	fluoride	5.2	96.06	sulphate	1.0
100.09	hardness as CaCO$_3$	1.0	32.06	sulphide	3.1
			58.08	thiocyanate	1.7
126.90	iodide	0.79	65.37	zinc	1.5

Table E

Determination of concentration by the known-addition method
The values were calculated from the Nernstian slope value, 59.16 mV

ΔE (mV)	C_x/C_Δ	ΔE (mV)	C_x/C_Δ	ΔE (mV)	C_x/C_Δ
1	25.20	11.0	1.87	18.6	0.941
1.5	16.23	11.2	1.83	18.8	0.927
2	12.35	11.4	1.79	19.0	0.913
2.5	9.79	11.6	1.75	19.2	0.900
3	8.07	11.8	1.72	19.4	0.887
3.5	6.85	12.0	1.68	19.6	0.874
4	5.94	12.2	1.65	19.8	0.861
4.5	5.22	12.4	1.61	20.0	0.849
5	4.65	12.6	1.58	20.2	0.837
5.2	4.46	12.8	1.55	20.4	0.825
5.4	4.28	13.0	1.52	20.6	0.813
5.6	4.11	13.2	1.49	20.8	0.802
5.8	3.95	13.4	1.46	21.0	0.791
6.0	3.80	13.6	1.43	21.2	0.780
6.2	3.66	13.8	1.41	21.4	0.769
6.4	3.54	14.0	1.38	21.6	0.759
6.6	3.41	14.2	1.36	21.8	0.748
6.8	3.30	14.4	1.33	22.0	0.738
7.0'	3.19	14.6	1.31	22.2	0.728
7.2	3.09	14.8	1.28	22.4	0.719
7.4	3.00	15.0	1.26	22.6	0.709
7.6	2.91	15.2	1.24	22.8	0.700
7.8	2.82	15.4	1.22	23.0	0.691
8.0	2.74	15.6	1.20	23.2	0.682
8.2	2.66	15.8	1.18	23.4	0.673
8.4	2.59	16.0	1.157	23.6	0.664
8.6	2.52	16.2	1.138	23.8	0.656
8.8	2.45	16.4	1.119	24.0	0.647
9.0	2.38	16.6	1.101	24.2	0.639
9.2	2.32	16.8	1.083	24.4	0.631
9.4	2.26	17.0	1.066	24.6	0.623
9.6	2.21	17.2	1.049	24.8	0.615
9.8	2.15	17.4	1.032	25.0	0.608
10.0	2.10	17.6	1.016	25.2	0.600
10.2	2.05	17.8	1.001	25.4	0.593
10.4	2.00	18.0	0.985	25.6	0.585
10.6	1.96	18.2	0.970	25.8	0.578
10.8	1.91	18.4	0.956	26.0	0.571

contin ued

continued

ΔE (mV)	C_x/C_Δ	ΔE (mV)	C_x/C_Δ	ΔE (mV)	C_x/C_Δ
26.2	0 564	29.6	0.462	37.5	0.303
26.4	0.557	29.8	0.457	38.0	0.295
26.6	0.551	30.0	0.452	38.5	0.288
26.8	0.544	30.5	0.439	39.0	0.281
27.0	0.538	31.0	0.427	39.5	0.274
27.2	0.531	31.5	0.415	40.0	0.267
27.4	0.525	32.0	0.404	41.0	0.254
27.6	0.519	32.5	0.393	42.0	0.242
27.8	0.513	33.0	0.383	43.0	0.231
28.0	0.507	33.5	0.373	44.0	0.220
28.2	0.501	34.0	0.363	45.0	0.210
28.4	0.495	34.5	0.353	46.0	0.200
28.6	0.489	35.0	0.344	47.0	0.191
28.8	0.484	35.5	0.335	48.0	0.183
29.0	0.478	36.0	0.327	49.0	0.174
29.2	0.473	36.5	0.318	50.0	0.187
29.4	0.467	37.0	0.310		

The values given hold for univalent ions; for bivalent ions, the values given correspond to the value $\Delta E/2$; e.g. C_x/C_Δ 25.20 corresponds to 1 mV for univalent and 0.5 mV for bivalent ions.

Table F

Determination of concentration by the known-subtraction method
The values were calculated by using the Nernstian slope value, 59.16 mV

$-\Delta E$ (mV)	C_x/C_Δ	$-\Delta E$ (mV)	C_x/C_Δ	$-\Delta E$ (mV)	C_x/C_Δ
1.0	26.196	11.0	2.871	18.6	1.941
1.5	17.633	11.2	2.830	18.8	1.927
2.0	13.352	11.4	2.791	19.0	1.913
2.5	10.785	11.6	2.752	19.2	1.900
3.0	9.074	11.8	2.715	19.4	1.887
3.5	7.852	12.0	2.680	19.6	1.874
4.0	6.936	12.2	2.645	19.8	1.861
4.5	5.224	12.4	2.612	20.0	1.849
5.0	5.654	12.6	2.580	20.2	1.837
5.2	5.458	12.8	2.549	20.4	1.825
5.4	5.275	13.0	2.518	20.6	1.813
5.6	5.106	13.2	2.490	20.8	1.802
5.8	4.949	13.4	2.461	21.0	1.791
6.0	4.802	13.6	2.433	21.2	1.780
6.2	4.664	13.8	2.406	21.4	1.769
6.4	4.535	14.0	2.380	21.6	1.759
6.6	4.414	14.2	2.355	21.8	1.748
6.8	4.300	14.4	2.331	22.0	1.738
7.0	4.193	14.6	2.307	22.2	1.728
7.2	4.091	14.8	2.284	22.4	1.719
7.4	3.996	15.0	2.261	22.6	1.709
7.6	3.905	15.2	2.239	22.8	1.700
7.8	3.819	15.4	2.218	23.0	1.691
8.0	3.738	15.6	2.197	23.2	1.682
8.2	3.660	15.8	2.177	23.4	1.673
8.4	3.586	16.0	2.157	23.6	1.664
8.6	3.515	16.2	2.138	23.8	1.656
8.8	3.448	16.4	2.119	24.0	1.647
9.0	3.384	16.6	2.101	24.2	1.639
9.2	3.322	16.8	2.083	24.4	1.631
9.4	3.264	17.0	2.066	24.6	1.623
9.6	3.207	17.2	2.049	24.8	1.615
9.8	3.153	17.4	2.033	25.0	1.608
10.0	3.102	17.6	2.016	25.2	1.600
10.2	3.052	17.8	2.001	25.4	1.595
10.4	3.004	18.0	1.985	25.6	1.585
10.6	2.958	18.2	1.970	25.8	1.578
10.8	2.914	18.4	1.956	26.0	1.571

continued

continued

$-\Delta E$ (mV)	C_x/C_Δ	$-\Delta E$ (mV)	C_x/C_Δ	$-\Delta E$ (mV)	C_x/C_Δ
26.2	1.564	29.6	1.462	37.5	1.303
26.4	1.557	29.8	1.457	38.0	1.295
26.6	1.551	30.0	1.452	38.5	1.287
26.8	1.544	30.5	1.439	39.0	1.281
27.0	1.538	31.0	1.427	39.5	1.274
27.2	1.531	31.5	1.415	40.0	1.267
27.4	1.525	32.0	1.404	41.0	1.254
27.6	1.519	32.5	1.393	42.0	1.242
27.8	1.513	33.0	1.383	43.0	1.231
28.0	1.507	33.5	1.373	44.0	1.220
28.2	1.501	34.0	1.363	45.0	1.210
28.4	1.495	34.5	1.353	46.0	1.200
28.6	1.489	35.0	1.344	47.0	1.191
28.8	1.484	35.5	1.335	48.0	1.183
29.0	1.478	36.0	1.327	49.0	1.174
29.2	1.473	36.5	1.318	50.0	1.167
29.4	1.467	37.0	1.310		

Table G

Determination of concentration by the analyte-addition method

ΔE (mV)	C_Δ/C_x	ΔE (mV)	C_Δ/C_x	ΔE (mV)	C_Δ/C_x	ΔE (mV)	C_Δ/C_x
1	0.0397	11.2	0.546	19.0	1.095	26.8	1.838
1.5	0.0601	11.4	0.558	19.2	1.111	27.0	1.860
2.0	0.0809	11.6	0.571	19.4	1.127	27.2	1.883
2.5	0.1022	11.8	0.583	19.6	1.144	27.4	1.905
3.0	0.1239	12.0	0.595	19.8	1.161	27.6	1.928
3.5	0.1459	12.2	0.608	20.0	1.178	27.8	1.951
4.0	0.1685	12.4	0.620	20.2	1.195	28.0	1.974
4.5	0.1914	12.6	0.633	20.4	1.212	28.2	1.997
5.0	0.2148	12.8	0.646	20.6	1.230	28.4	2.020
5.2	0.224	13.0	0.659	20.8	1.247	28.6	2.044
5.4	0.234	13.2	0.672	21.0	1.264	28.8	2.068
5.6	0.244	13.4	0.685	21.2	1.282	29.0	2.092
5.8	0.253	13.6	0.698	21.4	1.300	29.2	2.116
6.0	0.263	13.8	0.711	21.6	1.318	29.4	2.140
6.2	0.273	14.0	0.724	21.8	1.336	29.6	2.165
6.4	0.283	14.2	0.738	22.0	1.354	29.8	2.189
6.6	0.293	14.4	0.751	22.2	1.373	30.0	2.240
6.8	0.303	14.6	0.765	22.4	1.391	30.5	2.278
7.0	0.313	14.8	0.779	22.6	1.410	31.0	2.342
7.2	0.323	15.0	0.793	22.8	1.429	31.5	2.408
7.4	0.334	15.2	0.807	23.0	1.448	32.0	2.475
7.6	0.344	15.4	0.821	23.2	1.467	32.5	2.543
7.8	0.355	15.6	0.835	23.4	1.486	33.0	2.613
8.0	0.365	15.8	0.849	23.6	1.506	33.5	2.684
8.2	0.376	16.0	0.864	23.8	1.525	34.0	2.756
8.4	0.387	16.2	0.879	24.0	1.545	34.5	2.830
8.6	0.398	16.4	0.893	24.2	1.565	35.0	2.905
8.8	0.408	16.6	0.908	24.4	1.585	35.5	2.982
9.0	0.419	16.8	0.923	24.6	1.605	36.0	3.060
9.2	0.431	17.0	0.938	24.8	1.625	36.5	3.140
9.4	0.442	17.2	0.953	25.0	1.646	37.0	3.221
9.6	0.453	17.4	0.968	25.2	1.667	37.5	3.304
9.8	0.464	17.6	0.984	25.4	1.687	38.0	3.389
10.0	0.476	17.8	0.999	25.6	1.708	38.5	3.475
10.2	0.487	18.0	1.014	25.8	1.730	39.0	3.563
10.4	0.499	18.2	1.031	26.0	1.751	39.5	3.652
10.6	0.511	18.4	1.047	26.2	1.772	40.0	3.744
10.8	0.522	18.6	1.063	26.4	1.794	41.0	3.932
11.0	0.534	18.8	1.079	26.6	1.816	42.0	4.128

continued

continued

ΔE (mV)	C_Δ/C_x	ΔE (mV)	C_Δ/C_x	ΔE (mV)	C_Δ/C_x	ΔE (mV)	C_Δ/C_x
43.0	4.331	45.0	4.763	47.0	5.230	49.0	5.734
44.0	4.543	46.0	4.992	48.0	5.477	50.0	6.001

The values given hold for univalent ions; for bivalent ions the values given correspond to the value $\Delta E/2$; e.g. C_Δ/C_x 0.0397 corresponds to 1 mV for univalent and 0.5 mV for bivalent ions.
The values were calculated by using the Nernstian slope value, 59.16 mV.

Table H

Determination of concentration by the analyte-subtraction method

$-\Delta E$ (mV)	C_x/C_Δ	$-\Delta E$ (mV)	C_x/C_Δ	$-\Delta E$ (mV)	C_x/C_Δ	$-\Delta E$ (mV)	C_x/C_Δ
1	0.038	7.6	0.256	11.8	0.368	16.0	0.464
1.5	0.057	7.8	0.262	12.0	0.373	16.2	0.468
2.0	0.075	8.0	0.268	12.2	0.378	16.4	0.472
2.5	0.093	8.2	0.273	12.4	0.383	16.6	0.476
3.0	0.110	8.4	0.279	12.6	0.388	16.8	0.480
3.5	0.127	8.6	0.284	12.8	0.392	17.0	0.484
4.0	0.144	8.8	0.290	13.0	0.397	17.2	0.488
4.5	0.161	9.0	0.296	13.2	0.402	17.4	0.492
5.0	0.177	9.2	0.301	13.4	0.406	17.6	0.496
5.2	0.183	9.4	0.306	13.6	0.411	17.8	0.500
5.4	0.190	9.6	0.312	13.8	0.416	18.0	0.504
5.6	0.196	9.8	0.317	14.0	0.420	18.2	0.508
5.8	0.202	10.0	0.322	14.2	0.425	18.4	0.511
6.0	0.208	10.2	0.328	14.4	0.429	18.6	0.515
6.2	0.214	10.4	0.333	14.6	0.433	18.8	0.519
6.4	0.220	10.6	0.338	14.8	0.438	19.0	0.523
6.6	0.227	10.8	0.343	15.0	0.442	19.2	0.526
6.8	0.233	11.0	0.348	15.2	0.447	19.4	0.530
7.0	0.238	11.2	0.353	15.4	0.451	19.6	0.534
7.2	0.244	11.4	0.358	15.6	0.455	19.8	0.537
7.4	0.250	11.6	0.363	15.8	0.459	20.0	0.541

continued

continued

$-\Delta E$ (mV)	C_x/C_Δ	$-\Delta E$ (mV)	C_x/C_Δ	$-\Delta E$ (mV)	C_x/C_Δ	$-\Delta E$ (mV)	C_x/C_Δ
20.2	0.544	24.2	0.610	28.2	0.666	35.5	0.749
20.4	0.548	24.4	0.613	28.4	0.669	36.0	0.754
20.6	0.551	24.6	0.616	28.6	0.671	36.5	0.758
20.8	0.555	24.8	0.619	28.8	0.674	37.0	0.763
21.0	0.558	25.0	0.622	29.0	0.677	37.5	0.768
21.2	0.562	25.2	0.625	29.2	0.679	38.0	0.772
21.4	0.565	25.4	0.628	29.4	0.682	38.5	0.777
21.6	0.569	25.6	0.631	29.6	0.684	39.0	0.781
21.8	0.572	25.8	0.634	29.8	0.686	39.5	0.785
22.0	0.575	26.0	0.636	30.0	0.689	40.0	0.789
22.2	0.579	26.2	0.639	30.5	0.695	41.0	0.797
22.4	0.582	26.4	0.642	31.0	0.701	42.0	0.805
22.6	0.585	26.6	0.645	31.5	0.707	43.0	0.812
22.8	0.588	26.8	0.648	32.0	0.712	44.0	0.820
23.0	0.591	27.0	0.650	32.5	0.718	45.0	0.826
23.2	0.595	27.2	0.653	33.0	0.723	46.0	0.833
23.4	0.598	27.4	0.656	33.5	0.729	47.0	0.839
23.6	0.601	27.6	0.658	34.0	0.734	48.0	0.846
23.8	0.604	27.8	0.661	34.5	0.739	49.0	0.851
24.0	0.607	28.0	0.664	35.0	0.744	50.0	0.857

The values given hold for univalent ions; for bivalent ions the values given correspond to the value $-\Delta E/2$; e.g. C_x/C_Δ 0.038 corresponds to 1 mV for univalent and 0.5 mV for bivalent ions.

The values were calculated by using the Nernstian slope value, 59.16 mV.

Table I

Concentration by double known-addition
[(Taken from *Orion Res. Newsletter*, **II**, Nos. 7, 8 (1970)] by permission
of Orion Research Inc.

R	C_x/C_s	R	C_x/C_s	R	C_x/C_s
1.270	0.100	1.520	0.694	1.670	1.598
1.280	0.113	1.525	0.714	1.675	1.643
1.290	0.126	1.530	0.735	1.680	1.691
1.300	0.140	1.535	0.756	1.685	1.738
1.310	0.154	1.540	0.778	1.690	1.787
1.320	0.170	1.545	0.801	1.695	1.840
1.330	0.186	1.550	0.823	1.700	1.894
1.340	0.203	1.555	0.847	1.705	1.948
1.350	0.221	1.560	0.870	1.710	2.006
1.360	0.240	1.565	0.896	1.715	2.066
1.370	0.260	1.570	0.920	1.720	2.126
1.380	0.280	1.575	0.946	1.725	2.190
1.390	0.302	1.580	0.973	1.730	2.256
1.400	0.325	1.585	1.000	1.735	2.326
1.410	0.349	1.590	1.029	1.740	2.397
1.420	0.373	1.595	1.056	1.745	2.470
1.430	0.399	1.600	1.086	1.750	2.549
1.440	0.427	1.605	1.116	1.755	2.629
1.450	0.455	1.610	1.147	1.760	2.711
1.460	0.485	1.615	1.179	1.765	2.801
1.470	0.516	1.620	1.213	1.770	2.892
1.475	0.532	1.625	1.245	1.775	2.985
1.480	0.548	1.630	1.280	1.780	3.088
1.485	0.565	1.635	1·315	1.785	3.193
1.490	0.582	1.640	1.353	1.790	3.301
1.495	0.600	1.645	1.391	1.795	3.416
1.500	0.618	1.650	1.430	1.800	3.536
1.505	0.637	1.655	1.469	1.805	3.664
1.510	0.655	1.660	1.510	1.810	3.797
1.515	0.675	1.665	1.554	1.815	3.939

·Table J

Concentration by single known-addition with slope determined by dilution
(using 1 : 1 dilution)

$R=\Delta E_2/\Delta E_{dil}$	C_x/C_Δ	$R=\Delta E_2/\Delta E_{dil}$	C_x/C_Δ	$R=\Delta E_2/\Delta E_{dil}$	C_x/C_Δ
0.02	71.643	0.70	1.602	1.36	0.638
0.04	35.573	0.72	1.545	1.38	0.624
0.06	23.550	0.74	1.492	1.40	0.610
0.08	17.540	0.76	1.442	1.42	0.597
0.10	13.934	0.78	1.394	1.44	0.584
0.12	11.531	0.80	1.349	1.46	0.571
0.14	9.814	0.82	1.307	1.48	0.559
0.16	8.527	0.84	1.266	1.50	0.547
0.18	7.526	0.86	1.227	1.52	0.535
0.20	6.726	0.88	1.190	1.54	0.524
0.22	6.072	0.90	1.155	1.56	0.513
0.24	5.526	0.92	1.121	1.58	0.503
0.26	5.066	0.94	1.089	1.60	0.492
0.28	4.670	0.96	1.058	1.62	0.482
0.30	4.327	0.98	1.028	1.64	0.473
0.32	4.026	1.00	1.000	1.66	0.463
0.34	3.763	1.02	0.973	1.68	0.454
0.36	3.528	1.04	0.947	1.70	0.445
0.38	3.319	1.06	0.922	1.72	0.436
0.40	3.130	1.08	0.898	1.74	0.427
0.42	2.959	1.10	0.875	1.76	0.419
0.44	2.805	1.12	0.852	1.78	0.411
0.46	2.663	1.14	0.831	1.80	0.403
0.48	2.533	1.16	0.810	1.82	0.395
0.50	2.415	1.18	0.790	1.84	0.388
0.52	2.305	1.20	0.771	1.86	0.380
0.54	2.203	1.22	0.752	1.88	0.373
0.56	2.109	1.24	0.734	1.90	0.366
0.58	2.021	1.26	0.717	1.92	0.359
0.60	1.939	1.28	0.700	1.94	0.353
0.62	1.863	1.30	0.684	1.96	0.346
0.64	1.791	1.32	0.668	1.98	0.340
0.66	1.724	1.34	0.653	2.00	0.333
0.68	1.661				

Table K

List of Abbreviations

Aliquat 336S — methyltricaprylammonium chloride
ChelDP — ethylenediaminedi(o-hydroxyphenyl)acetic acid
DCTA — 1,2-diaminocyclohexanetetra-acetic acid
EDTA — ethylenediaminetetra-acetic acid
EGTA — ethyleneglycol-bis-(2-aminoethylether)tetra-acetic acid
ISE — ion-selective electrode
LISE — liquid ion-selective electrode
NTA — nitrilotriacetic acid
TEPA — tetraethylenepentamine
Tris-buffer — tris(hydroxymethyl)aminomethane

Index